"We cannot afford to ignore ESP...

"So far, the purely physical sciences have studied only the material components of human life—our chemistry, anatomy and physiology. Yet we know there is much, much more to a man than atoms, molecules, blood, bone and muscle. To say that the human spectrum ends with the portion we directly experience is like saying the universe ends where it ceases to be visible.

"It is this unknown end of the human spectrum that parapsychology—the frontier science of the mind—is probing. It is in these probings that science and the human spirit may find their ultimate meeting place. No one who is informed on the subject of ESP can disregard it and still remain honest with himself."
—HENRY PIERCE

Other SIGNET Titles of Related Interest

☐ **THE PSYCHIC WORLD AROUND US by Long John Nebel** with Sanford Teller. Long John Nebel, America's leading nighttime-radio personality, recounts his experiences with psychic phenomena. (#Q4201—95¢)

☐ **A PSYCHIATRIST LOOKS AT ESP (Psychic Dynamics) by Berthold Eric Schwarz.** A distinguished psychiatrist presents an enlightening study of the nature of psychic phenomena in people who are psychically gifted. (#T3570—75¢)

☐ **YOUR MYSTERIOUS POWERS OF ESP by Harold Sherman.** Extraordinary cases involving extrasensory perception and ways to release your own mysterious powers of ESP are revealed. (#T4104—75¢)

☐ **TRUE EXPERIENCES IN EXOTIC ESP edited by Martin Ebon.** An extraordinary digest of some of the strangest happenings in the world of the supernatural—cases of death by spell, magical healings, spirit exorcism, walking on fire, and other phenomena. (#P3363—60¢)

THE NEW AMERICAN LIBRARY, INC., P.O. Box 2310, Grand Central Station, New York, New York 10017

Please send me the SIGNET BOOKS I have checked above. I am enclosing $_____(check or money order—no currency or C.O.D.'s). Please include the list price plus 10¢ a copy to cover mailing costs. (New York City residents add 6% Sales Tax. Other New York State residents add 3% plus any local sales or use taxes.)

Name_____

Address_____

City_____State_____Zip Code_____

Allow at least 3 weeks for delivery

Science Looks at ESP

HENRY W. PIERCE

A SIGNET MYSTIC BOOK from
NEW AMERICAN LIBRARY
TIMES MIRROR

To my wife, Celeste

Copyright © 1970 by Henry W. Pierce

All rights reserved

SIGNET TRADEMARK REG. U.S. PAT. OFF. AND FOREIGN COUNTRIES
REGISTERED TRADEMARK—MARCA REGISTRADA
HECHO EN CHICAGO, U.S.A.

Signet, Signet Classics, Mentor and Plume Books
are published by The New American Library, Inc.,
1301 Avenue of the Americas, New York, New York 10019

First Printing, June, 1970

PRINTED IN THE UNITED STATES OF AMERICA

Contents

I	Mysterious Events	7
II	"Nothing But" Man	21
III	The Search for a Repeatable Experiment	32
IV	Telepathy and Clairvoyance	44
V	Seeing the Future	57
VI	Hypnosis and ESP	70
VII	Mind over Matter	81
VIII	Mind over Life	93
IX	Parapsychology's Far-out Frontiers	101
X	Will Russia Score First?	125
XI	Other Science Writers View ESP	135
XII	Summing Up	140
	Selected Bibliography	143

* 1 *

Mysterious Events

If you told me your soul once left your body, I would be strongly inclined to call you a crackpot. We all know such things simply don't happen. And anyone who says they do is, by all the standards of 20th-century American society, a likely candidate for a mental hospital.

The very idea that a person's mind could somehow function outside his body is a throwback to the darkest of the dark ages of superstition. Yet I have met people who insist that precisely this kind of thing has happened to them. They visit me occasionally in the city room of the Pittsburgh *Post-Gazette,* where I report the news of the day in the fields of science and medicine. One fellow, a tall, blondish, not-too-well-educated man in his early 40's, claimed he saw and visited with his dead wife in his dreams. He regularly left his body also, he said, visited various parts of his house and his neighborhood in a disembodied state, and wrote poetry while drifting about in the immaterial. It was the poetry he wanted me to evaluate— and, of course, publish in the *Post-Gazette*.

I don't have much time for characters like this. Typically, they offer no evidence of any kind to back up their claims. If you suggest a test—some simple little thing like drifting over to your house, making a note of what's on your dining room table, and reporting the experience to you the next day—they invariably get angry. (My tall, blondish friend snapped up his poetry, slammed the briefcase shut, and stalked off in a huff.)

So it is with some reluctance that I tell the following

story. I relate it in the full knowledge that, after it has been published, I shall be subject to every sort of ridicule that men of common sense can impose. I know. I have been among the loudest scoffers.

It happened on the night of May 21, 1966. My wife Celeste and I had gone to bed about 10 p.m. Roughly an hour later I woke up with the strong impression that I had somehow been communicating with her. I had the feeling —a certainty, really—of having been in contact with Celeste while we were asleep, and that, somehow, this contact had something to do with her health. She had been in the best of health when we went to bed, yet for some reason—which I was wholly at a loss to explain—this sense of having been in contact with her was accompanied by strong anxiety over her physical condition. As I was waking up, I found myself asking her, "Are you all right?"

But there was a stranger part still. For a split-second while waking, I also had the feeling of having been partly out of my body, and of re-entering it or taking control of it again. I can't blame you if you smile at this. As I sit here re-reading it, it sounds impossible even to me—yet I recorded it as faithfully as I could within 20 minutes of the time it happened. It was as if I had momentarily abandoned control of my body, rather like unplugging myself from it or like disconnecting the circuits—as if my mind had been needed to do some other job.

I was waking, then, with this sense of having been communicating with Celeste, concerned over her health, and of now re-entering my body. As I "took control" of my body, I had a feeling—very briefly—much like that of slipping back into a sweater or suit that I had partly slipped off; I felt, in a sense, like a knife that had been pulled part-way out of its sheath and was in the process of slipping back into it again.

I was firmly convinced that Celeste had shared a portion of the experience with me. I knew—that's the only word for it—that she could tell me what had happened if I asked her immediately. So I did.

She, too, was just waking up. She replied that she had been lying in bed, in a state between sleeping and waking. She suddenly felt an irritating sensation in her throat, as if she were about to choke. She was afraid the choking would be fatal. "I was too frightened to call out or make the slightest sound—I was afraid the least effort to speak would kill me," she said later.

Then, quite suddenly, she sensed what appeared to be

my spirit, or mind, or whatever you care to call it, near her head. "I couldn't actually *see* you physically, yet I knew it was you," she told me. "Yet your body," she said, "was still on your bed."

My own twin bed, some three feet away from hers, held my physical self while my wife experienced some other "me" near her head—which matched my own experience of the event beautifully.

And yet another strange thing had happened. Somehow, she said, I had used my mind to relieve her choking. "I don't know what you did, but whatever it was, it was quickly done," she declared.

I went downstairs and wrote out a full description of the entire event. It was 11:20 p.m.

The following morning Celeste and I discussed the experience in more detail. Here is her account:

"I was panicked that I was going to choke. Then all of a sudden you did something with your mind, and I had the feeling you had completely dispelled the choking and had saved me. It was quickly done. I wasn't sure how you had done it. It was some kind of strange force. I was aware of lying in my own bed and of you lying in your bed, but it seemed as if you were also away from your bed and close to me—you were on your bed but you were also off of it, part of you had left it—I can't explain it. But you very quickly did away with the choking.

"Right after this you started asking me if I was all right. That seemed like a natural question, but it also seemed like a strange thing to ask—you knew I was all right because you had saved me. Your next question was, 'Why did I ask you that?' and it brought full consciousness—it startled me because it was such a ridiculous question under the circumstances."

She said she was under the impression that she was awake throughout the experience until, paradoxically, she found herself waking up. She later described her state as "between sleeping and waking." The room we had appeared to be in, for her, she said, wasn't quite like our "real" bedroom—it was more like a hospital room, and our beds were at right angles to each other instead of parallel.

She was puzzled over what had seemed to be different levels of consciousness. She expressed bewilderment over the problem of putting certain parts of the experience into words. While she was waking up, she said, she had trouble finding words to fit an "other dimensional" quality of the experience; and as she woke more fully and began to gain more complete control over her verbal processes, she found the "other dimensional aspect" of it slipping from her memory. It was almost as if the two levels of consciousness were mutually incompatible—as if the three-dimensional concepts we depend on in "normal" awareness do not apply on this other level, and that, given one, the other must go.

I can not explain this experience. I am not offering it as proof of any philosophy or any religion or any theory. I am simply reporting what happened.

Celeste and I had one other mysterious experience worth noting. It happened shortly after we were married. We had gone to bed at our usual hour—about 10 p.m. or shortly thereafter. I woke out of a deep sleep at about 2 a.m., unaccountably angry with Celeste and very worried about my dog, which was staying with my parents in another city. I was so worried about the dog that, despite the hour, I almost got up to call my parents long distance. The next morning Celeste said she had dreamed that, since the dog was old and unwell, she and my mother had taken him to a veterinarian to be "put away." She had done this, she said, without consulting me—an act that would certainly have aroused my anger had it been done in real life. Did her dream somehow cause my anger toward her and my anxiety about the dog? I do not know.

Such experiences, of course, are not peculiar to Celeste and me. As a matter of fact, they may be much more common than most of us suspect. Nearly everyone has had the experience of thinking about another person, only to hear the phone ring and, on answering it, to discover that the caller was the very person who had occupied his thoughts.

On the night of April 16, 1969, a woman of my acquaintance was sitting with her husband in their bedroom, conversing before bed. The time was 11:55 p.m. Suddenly she heard the phone ring.

"I asked my husband to answer it," she told me the next day, "but he insisted it wasn't ringing. I knew better, of course—I had heard it. It continued to ring. But when he finally answered it all he got was a dial tone."

Five minutes later, she said, the same thing happened: she heard it ring, her husband did not, and when the phone was answered—this time by the woman—there was only the familiar dial tone.

Five minutes later the phone rang again, but this time

both of them heard it. He answered it. The caller was a mutual friend who had recently moved out of town and, after leaving the city, had lost his wife in an auto accident. He was calling long distance.

For the previous ten minutes, he told them, he had been on the verge of phoning them several times but had hung the phone up each time before completing the call.

Coincidence?

Some people seem much more prone to these experiences than others. I've known husbands and wives who, night after night, were seemingly able to read each other's thoughts—spontaneously and uncontrollably, and as if the thoughts of the other were able to intrude only when they were least expected. In one such case, the husband came home bearing news of the pregnancy of a mutual friend, whom I will call Jane.

"Guess what?" he said to his wife as he burst in the door.

Without further ado, his wife replied: "Jane's pregnant."

She hadn't been told of Jane's condition. She hadn't seen Jane recently. It was just a hunch. It came suddenly. She blurted it out, almost on impulse. Such things happen often to this couple.

I can't remember exactly when I first became interested in psychical research. Perhaps it was those ghost stories we used to tell each other as children. (I know at least two people who became chemists because of an early interest in stage magic!) Perhaps it was a story told me by an aunt, of seeing an apparition of a man sitting beside her bed—a man in white. Perhaps it was a dream I had one night at the age of about 11—a dream of a skeleton looking at me through my bedroom window—which was followed by the discovery, the next morning, that a burglar had reached in the window of my parents' bedroom, opposite mine, and had stolen my father's trousers from a chair near his window. But if any of these early experiences did influence me, they influenced me only in the most tangential way: I didn't even learn there was such a thing as psychical research, as a formal study, until much later.

Not so, however, with my interest in science. Shortly after I was born, my mother told my father that, if she had one wish for me and my future values, it was that I would get enjoyment from simple things. My father, a research chemist, replied that his wish was that I would enjoy science.

Both their wishes were realized, but my father's had plenty of help from the rows and rows of books on science and philosophy and other subjects that adorned our book shelves, from a fascinating electric motor that was kept tantalizingly out of reach, and from his own habit of throwing out intriguing bits of information and fascinating questions.

"If this wallpaper were up in the sky about sunset," he ventured one night, gesturing toward a vast desert of dirty yellow wallpaper that graced a friend's dining room, "we'd probably think it was quite beautiful. But here in this room it is not." Truth, apparently, was relative.

And those books! What marvelous secrets they held, with their unforgettable pictures of flowing galaxies and spiral nebulae, the imaginative renditions of moonscapes and Martian landscapes, and the careful artists' portrayals of fearsome dinosaurs and mysterious ancient seas! I yearned toward science.

But alas, my math gave me problems. To this day I can remember my second-grade teacher laboring to explain how you took one from seven and got six. Math would certainly have been my undoing had not my initial scholarly failing there been redeemed by a modest aptitude for writing and spelling.

So, although I retained a deep love for science and for the scientific spirit throughout my school years, I found myself shifting ever more toward a career in journalism. Yet I had no thought of becoming a science writer until my junior year at the University of Wisconsin School of Journalism.

It happened this way. The University kept a confidential file on its journalism students. Each student was rated by his professors in terms of his interests, abilities, personality, diligence, and so on. When I found out about this file I was seized by a burning curiosity and not a little resentment. What business did they have keeping a secret file on me, anyway? I resolved to see it the first chance I got.

I never really expected to get such a chance, of course. But, by the sheerest luck, I did. I had stopped in at the journalism office to attend to some minor affair, when I noticed a student record sheet sitting on the counter in plain sight. A glance told me that this must be one of those confidential reports—it carried little items like "Confidence," "Personal Attractiveness," and "Special Interests."

And, surprise of surprises, the student whose traits this sheet recorded was none other than myself!

A secretary saw me scrutinizing it and lost no time in whisking it from view—but not before I had seen this notation, entered by an instructor in feature writing:

"Given his choice, he writes on science subjects."

I hadn't been aware of it. But there it was. And, for me, it was the beginning of a career.

Despite my respect for science, I became disillusioned with much that passes for scholarship—especially in the social sciences—fairly early in my college career. A course in freshman economics made me a thoroughgoing cynic. The textbook was a conglomeration of unwieldy sentences concocted by someone who either didn't know how to write or was trying mightily to impress less learned persons with his scholarship. The ideas were basically simple, yet they were expressed in impossibly complex terminology. In my journalism courses I was being trained to write simply. Short, declarative sentences were the order of the day. So I tried applying my journalism training in this economics class. Each Friday we were given brief written quizzes on the week's work in economics, and I endeavored to phrase my answers in the simple, straightforward style of newspaper journalism. And each week I was given a failing grade.

But one Friday we were given a question which (1) I couldn't read and (2) even if I had been able to read it, I doubted whether I would have known the answer. It didn't look like anything that had been buried in those impossible sentences in the textbook. In short, I didn't have a hare's idea of what the question was all about.

Desperate, I started inventing words and phrases. Words like "hedinical," which aren't in any dictionary. I strung these out in fine, pompous sentences that meant absolutely nothing. I threw in a phony fact or two for good measure. When it was done, the essay was composed of sentences that sounded something like this:

"There is no retrogressively apparent causative element in formulae of the manifest nature exhibited herein, save for two instances of external dominance in the steel industry in 1927."

I expected—and should have gotten—a swift kick in the pants from my instructor. Instead he gave me a perfect grade—no changes, no corrections. Great job, Pierce!

Thus rewarded, I pursued my evil ways. Each week my answers grew more meaningless, and each week I received perfect grades. Eventually my instructor called me in to congratulate me on my fine performance.

I might have written it all off as the stupidity of one lousy instructor were it not for one fact: by design, the students' papers were rotated among five different instructors. In other words, a different instructor must have been taken in by my chicanery each week.

I learned a lesson from this. It was a bitter, cynical lesson. But it has proved itself time and again since. There is a great deal of phony, hollow, blown-up pomposity in many so-called "respectable" scholarly papers. Furthermore, if you weed out the jargon, simplify the sentences, and state the basic ideas in good common English, you end up with some pretty self-evident stuff—statements like, "When the thermometer rises the temperature is going up." Or, "When a lot of people want something that happens to be scarce, its price will go up."

I have seen this kind of thing over and over in my journalistic career. I recall a professor at the New York State College of Agriculture applying for a federal grant to study a certain insect pest that was ruining apples. In writing his application, he called an apple an apple and gave the pest a fairly common name. His request was turned down. He expressed his disappointment to a more sophisticated colleague, who took a look at his application and said, in effect, "No, old boy, we don't do it like that. We do it like this." And he re-wrote the application, using all the appropriate jargon. Sure enough, the re-written application was approved, and the man got his money.

But the lesson went deeper than this. Scholarship, I discovered, was not entirely a search for the truth. It had its share of status seeking. It, too, was pressured to conform. It had its "in" groups and its "out" groups. It had subject matter that was fashionable and other subject matter that was unfashionable. It had its symbols of authority.

This helped me to understand, later, why so many scholars scoff at psychical research.

It also helped me to understand why psychical research is relatively free of the jargon that infests the social sciences: psychical researchers are not trying to impress the academic world with their scholarly abilities—if they were, they wouldn't be doing psychical research. Furthermore, the material the best ones study is complicated enough without the added burden of frothy, frilly gobbledygook.

My college career was interrupted by a two-year hitch in the army, during which I saw service in Korea and Japan. After I returned, I finished college, worked for a short time for a science news syndicate in Washington, D.C., and then, having earned college benefits under the G.I. Bill, I decided to study for a master's degree in Psychology—a subject that had always interested me.

I chose the University of Montana, partly because of its superb setting in the Rocky Mountains. During my graduate training at Montana I was exposed to a wide variety of views and "systems" of psychology, from the orthodox schools of psychoanalysis to behaviorism, neo-behaviorism, and Gestalt psychology. I learned to use hypnosis—which was essential to the experiment I carried out for my master's dissertation. And, for the first time, I discovered the feeling of warmth that comes with having professors who will defend your right to undertake controversial research—even help you with that research—although they themselves do not agree with the premises on which that research rests.

I received my M.A. in psychology in June, 1957.

For a time after that, I seriously considered going into psychology professionally. I worked at two New York State mental institutions for one year, chiefly giving psychological tests. But I found myself becoming increasingly disenchanted with the field. For one thing, I felt there were a great many questions about the nature of the human mind which psychologists were ducking. It seemed to me then (and still seems) that many of the most pressing questions about consciousness and human emotions were being avoided by psychologists who were over-eager to have their profession recognized as a full-fledged science. The way to obtain such recognition, they seemed to feel, was by ignoring all those facets of human nature which do not lend themselves to study by the methods of the physical sciences.

For another thing, I wanted to go on writing articles, putting my journalism training to some use. But I found that this was frowned on in some of psychology's professional circles. I remember the horror with which the head psychiatrist greeted the news that I was a writer. I had innocently been taking notes on an address he had been giving the staff. Then, when I contacted him after his talk and asked for clarification of a few points—for a possible magazine article—he became grossly ill-at-ease and even expressed a fear that I might write an "exposé" of his institution!

In 1962 I joined the staff of the *Post-Gazette* as science writer and columnist. I have found greater freedom to write the truth as I see it—to "tell it like it is"—in newspaper work than in any other capacity I have ever filled. Newspaper people discipline themselves to think and write in the simplest possible terms. Frills and froth are stripped away. So, too, is a great deal of sham and pretense. Facts are reported on their own merit. If an event is known to have occurred, you can write about it. No one tells you not to write it because it's "impossible."

My interest in psychical phenomena had been slowly growing throughout these years. Originally, perhaps because of my early identification with science, I had scoffed at much of it. I still do. But I have found that, despite my initial skepticism, there is a hard core of material that defies explanation.

You have to search diligently, with an open but critical mind, to find it, however. A considerable mass of material in this area comes across the desks of reporters. Most of it isn't worth spending time on. But a few cases—a very few indeed—are genuine and worth investigating.

The first case I ever investigated personally, on the scene and in detail, involved a professor from a Pittsburgh university who lived in a house he thought was haunted. He and his wife were two-feet-on-the-ground, unsuperstitious types. Neither of them believed in ghosts. Certainly they had never expected to see one.

To protect their identities I will call him Frank and her Nora. Early in the 1960's Frank, his wife, and their two children moved into a 200-year-old house in western Pennsylvania. They had lived there several months, Frank told me, when his wife noticed that a door between the parlor and the cellar had a mysterious tendency to swing open—apparently by itself. Nora was certain the door was usually securely shut. On one occasion, she claimed, she got a clear view of the door handle turning of its own accord before the door swung inward.

Nora was puzzled, but Frank considered it simply a matter of wind, or gravity, or some bit of perversity in the structure of the house itself. Slight contractions might take place in the door that would cause the latch to come loose, making the handle turn a bit and the door swing open. Certainly no cause for excitement in that.

As the obstinate door continued its unorthodox behavior, however, Nora did start to get a bit apprehensive. Perhaps the age of the house and the somewhat rural setting combined to impose that commonly felt spookiness that everyone must know who has ever visited an old building at night. Nothing of any consequence happened for several more months. Then she saw *it*. . . . But, surprisingly, it didn't frighten her. It gave her a warm, comforting feeling.

The time was about 11 p.m. Nora had gone upstairs to take a final look at the baby before retiring—a practice she and her husband followed each night. She had no sooner entered the baby's room than, after a brief glance at the child's crib, she saw what she later described as an old woman sitting in a chair near the child, rocking. The woman was gray-haired, she said, wore full skirts and lisle stockings. The room was illuminated by light coming in from the hall and by moonlight that entered through the lone bedroom window.

Afraid Frank would scoff at her—he had roundly pooh-poohed her apprehensions over the door incident—Nora did not mention the apparition to him. But she saw it again the following night.

The night after that her husband went up to check on the baby.

He, too, saw it.

But, being the skeptical member of the family—the discounter, the explainer-away, so to speak—he was hardly in a position to go downstairs and tell Nora he had seen a ghost.

He didn't mention it.

A night or two later he saw it again. Again, he went downstairs without mentioning it. But his face betrayed him. "Did anything seem unusual to you upstairs?" Nora asked.

Frank asked her what she meant. She said only that she had seen something unusual.

At that point Frank took a precaution which, because it is so rarely employed under these circumstances, makes this whole event outstanding. He proposed that, instead of telling each other what they had experienced, they sit down and write out separate descriptions of it in as much detail as possible, then compare notes.

Their descriptions tallied closely. They described her as being short, having gray hair drawn back. Nora's account included more details of her clothing. As Nora described her, her hair was tied in a bun at the back, she had a sweater that buttoned down the front, a full skirt, and lisle stockings. Shoes were not noted. Both Nora and Frank said she was "all in grays." Neither was able to identify the clothes as to period.

A second "haunting" reportedly occurred in the same house. It happened after Nora, sitting in the living room one night, saw the handle lift on that cellar door and the door slowly swing open. She told her older daughter to close the door, which the girl obediently did. The handle lifted again, and the door opened. This happened two or three times, Nora said. She told Frank about it, but he discounted it—in effect, he told her the house was "getting to" her. Then, several nights later, Frank saw the same thing—the handle apparently glittered as it moved, catching his eye. The door opened. When Frank went to close it he saw a figure standing at the bottom of the cellar stairs.

Frank was frightened. The only other entrance to the cellar—through the garage—was locked. As Frank described the figure, it was not like the sort of clear visual image you have when you see a normal living person. Frank said he somehow "felt" it as much as "saw" it. It was a man, he said, and he got the impression that the man was wearing a shirt open at the neck and had one foot on the stairs, looking up. Frank shut the door quickly. The next day he put a new lock on the door, and since then a bookcase has been built in front of it. His feelings about the man, he said, were distinctly unpleasant —in contrast to his feelings about the old woman figure.

I heard about Frank and Nora's experience through a mutual friend. I had never met them, but the account interested me sufficiently so that, very soon after I received this second-hand report, I called the professor, introduced myself, and asked if I might visit him and discuss the alleged apparition.

He agreed, and on Friday, March 5, 1965, I trod up the unpaved muddy driveway to his house.

Neither he nor his wife struck me as being "crackpots" in any sense. They were Presbyterians, not strongly religious. They seemed to take a very casual attitude toward the apparition. "We try to regard it as something to laugh at—a sort of joke," he declared. He confessed, however, that part of this light-hearted attitude might be a way of denying the whole thing to themselves: "We sort of wish it hadn't happened and we don't want to take it too seriously," he commented. The experience, he said, had already caused the family embarrassment: one of their small daughters had told her schoolmates about it, and Frank subsequently received an irate call from a parent protesting what she called "Frank's ghost stories."

"I wish people wouldn't take it seriously—I wish they'd just leave us alone," Frank said later.

I talked with Frank and Nora separately at first, obtaining independent versions of the affair. At my request, Frank and Nora led me, step by step, through the entire sequence of events, re-enacting the whole affair as far as possible.

The baby's bedroom measured approximately eight by ten feet. The chair was no more than two feet from the window, at about a 45-degree angle to it. There was only one entrance to the bedroom.

Frank and Nora said that, from the time they moved in, the room seemed to have a feeling of warmth, peace, and light about it. A short time after the apparition began appearing, Frank and Nora had the room redecorated. After they did that, they said, two things happened: the room lost its special warmth, and the figure stopped appearing.

Who or what was this apparition? Was it some eerie illusion produced by moonlight? A hallucination shared by both husband and wife, at different times? Had their minds unexplainably leaped back to some past event, perceiving, retrocognitively, a long-dead person? Was the whole thing a hoax?

A neighbor who had been told of their experience expressed the view that the apparition might "be" an aunt of an elderly friend of hers who had lived there many years before. The aunt, reportedly, was very fond of children and enjoyed being near them. There is nothing to substantiate this, however.

The important thing about this story is that, if it is true, it represents one of the few cases in which people have seen an apparition at different times and written down their descriptions of it before telling each other what they saw. Unfortunately, however, they did not save the written descriptions—as in so many cases of this type, we have to take their word for it.

Such experiences represent an entire class of events that the present scientific philosophy can not explain. If you accept the view of the universe put forth by most scientists today, you must conclude that people who tell such stories are either lying or badly deluded. There simply is no other way the more striking of these occurrences can be handled in terms of today's scientific traditions. Coincidence can account for some of the less spectacular happenings; some cases can also be explained in terms of subtle, forgotten, or overlooked bits of communication between people.

But for the weirder, more far-out cases, this simply will not suffice. The only way these alleged happenings can be dealt with, in the dominant scientific tradition, is by denying that they ever took place at all—or at least by denying that they happened as reported. If they did happen, we must either change our conception of nature, or admit that our conception, while correct, represents a much smaller part of reality than we have suspected. In either case, the consequences for religion and philosophy would be stupendous.

* II *

"Nothing But" Man

Ever get that blue, droopy mood of not caring? Know how it is to feel utterly meaningless, as if life didn't count? Sure you do. Everybody does. A certain amount of it is natural. But it can get too frequent. It can dominate your life. When it does, you're sick.

I believe many people are sick in precisely this way. I think that's partly why we're sometimes called a "sick society." I believe the sickness has its roots in the extremely low opinion we have of ourselves—a low opinion of ourselves collectively, as a species, fostered partly by a disappearance of the old church-related values.

True, many people still go to church. But even for them, of course, many of the old values don't necessarily apply. An example, if I may: once some friends invited my wife and me to an overnight social affair sponsored by their church. That night a young unmarried Sunday School teacher, who had come in the company of a young man, left the cot in which she had been sleeping and joined another young man whom she had met only a few hours before. They proceeded to have sexual relations on a cot about 10 feet from my wife and me. The room was quite dark and perhaps they thought all the other people in it were asleep. I am not passing a moral judgment on this young couple—that's not my purpose. What they did may have been perfectly all right—or it may not. The point is, being a member of a church does not, in itself, mean that an individual holds to traditional guidelines and beliefs, even within the social confines, as it were, of the church's society.

Part of the trouble stems from our de-humanization at the hands of our own science and technology. We have become, as others have said, I'm aware, mere numbers, in a sense—digits in a big machine. We are victims of our vastly complicated society and our complicated, impersonal gadgetry. In many ways, we are holes in punch-cards being fed into a gigantic computer.

But the trouble goes even deeper. It's not just that society is big and we are small. We have lost our faith in our very souls. We have been too willing to abandon our beliefs in our own spiritual nature—too willing to slaughter these beliefs on the altar of science long before the facts warranted such a sacrifice.

For 300 years we have been undergoing a slow, steady debasement at the hands of some philosophers who eagerly over-generalized each new discovery about mankind, stretching each into a theory that explained man too simply. Over and over we have been told we are "nothing but" dust, or "nothing but" an inexpensive collection of chemicals, "nothing but" a behavioral mechanism, until at last we have begun to believe it. Too many of us view ourselves, quite literally, as "nothing but" men. Is it so surprising, then, that we have lost our regard for ourselves and our fellow humans? Is our loss of purpose really so unexpected?

I am not deploring science as such. Nor am I opposed to the search for truth. Quite the contrary. I am urging that the truth be sought. What I am opposed to is the use some scholars make of too few facts—their readiness to deny man a mind or will because these have no place in physics. I resent being told that I am "nothing but" a chemical complex simply because the more spiritual parts of me won't fit into the test tube.

Our ancient sense of self-importance, we are told, was based on myths and superstitions. Through most of our civilized history we have been bemused with the notion that we were God's specially created beings. The world, we thought, had been made especially for us. Even in the darkest hours of tribalism, when we stood helpless before nature's tempests, we still flattered ourselves that those tempests had been summoned on our account—summoned, perhaps, to punish us for our trespasses against the gods.

Childlike, we clung to our gods even when they visited their wrath upon us. For these very gods gave us a sense of importance. And was it not better to feel important—though afraid—than to feel that the whole human race was an accidental, meaningless affair? We were like small boys and girls who would rather have their parents punish them than ignore them.

We spun ourselves a wide variety of religious beliefs. Some of us believed in a few gods, some in many. Some of us believed the afterworld was a shadowy, unimportant place. Some of us deified the sun, some the moon, some rain and thunder. Some of us worshiped our ancestors. Some invoked spirits. But through all our diversity we held one belief in common: we considered ourselves inherently different from and superior to the material universe.

We were, we thought, gifted with at least a certain amount of free will—and with it went moral responsibility. There was "wrong" and there was "right" and we flattered ourselves that we had been granted the power to choose one or the other. From such notions were developed concepts of honor, honesty, and integrity—moral yardsticks by which all men might be measured.

Then began what might be called the Great Erosion— the steady wearing away of our self-esteem. Astronomy unseated us from our imagined throne at the heart of the universe. Biology rudely thrust us from our privileged position among living beings—we took our places with the apes and cattle. Anthropology showed us we were no more intelligent—and certainly no better morally—than our cave-dwelling ancestors. Psychology, spearheaded by Sigmund Freud, excavated the more odious contents of our unconscious selves and told us that this, not reason, underlay most of our so-called "rational" behavior.

By the 1920's philosopher Bertrand Russell was able to tell a disillusioned world. . . .

> That man is the end product of causes which had no pre-vision of the end they were achieving; that his origin, his growth, his hopes and fears, his loves and his beliefs, are but the outcome of accidental collocations of atoms; that no fire, no heroism, no intensity of thought and feeling, can preserve individual life beyond the grave; that all the labours of the ages, all the devotion, all the inspiration, all the noonday brightness of human genius, are destined to extinction in the vast death of the solar system, and that the whole temple of man's achievement must inevitably be buried beneath the debris of a universe in ruins— all these things, if not quite beyond dispute, are yet so nearly certain that no philosophy which rejects them

can hope to stand. . . . Such in outline but even more purposeless, more void of meaning is the world which science presents to our belief.*

What was most surprising, perhaps, was that men like Bertrand Russell were able to lead dedicated, useful lives despite their self-proclaimed beliefs in the purposelessness of the universe. One wonders whether they were really as convinced of man's dusty origins as they said they were. After all, we note, many believers in "nothing but" man are among the loudest spokesmen for world peace. Some experts who proclaim materialism most loudly are among those most concerned lest man be incinerated in a very material war.

These people, perhaps, either do not believe their own philosophies, or are able to find some extenuating circumstance that justifies the existence of a meaningless man, or are so involved in the search for truth that the impact of materialism escapes them. But what happens when materialism creeps down from the towers of academia into the ranks of common men? What happens when our intellectual and spiritual leaders—our teachers and many of our clergymen—can no longer seriously promote an ideal image of man? What happens when our schools can no longer tell our children that man is a sacred being with a sacred destiny?

It is not simply a matter of reading the Bible in the classroom—an exercise which, I suspect, never carried much meaning anyway. It *is* a matter of being able to transmit, through words and gestures, a firm belief in honor and justice. The problem is that we should be able to say something like this to our children:

"This kind of behavior is right, and that is wrong, whether you and I like it or not. Right and wrong go beyond us. They are absolute. They have roots in something greater than mankind. We did not invent them. They existed before we did and they will exist after we are gone. When you rebel against them you are not rebelling against standards created by adults; you are rebelling against something that exists in itself, something you can not change."

But a materialistic universe has no place for such things. Indeed, it has become intellectually stylish to de-

* So quoted in Langdon-Davies, John, *On the Nature of Man*, The New American Library, 1960, p. 11.

bunk any belief that might lend dignity to the human race. Thus truth-telling has gone out of style, and fidelity has become a quaint practice which our children class with Amish buggies. Thus honor and justice have become jokes —why bother with justice for "nothing but" man?

Materialism has become, then, much more than a mere framework in which to view nature. It has become much more than an academic method, much more than a way of looking at the laws of the universe. The same philosophy which says "free will is an illusion and spirit is a myth" must also say "the meaning of life lies in the personal pleasures you can obtain from it."

Hence the rising crime rate. Hence our indifference when a fellow human is attacked in plain sight on a street corner. Hence our emphasis on non-involvement—our studied avoidance of conduct which, though "right" by an older standard, might subject us to controversy.

Material goods are the only things worth striving for, to most of us, because they are all we really believe in. This is materialism—the same materialism which pervades science, but the other side of the coin. It is materialism manifesting itself as an ethic rather than as a way of viewing the universe scientifically. How can one expect our young people—or anyone else—to be committed to anything beyond themselves when our materialistically based philosophy teaches that there is nothing outside our personal gratifications for which to live? It is pointless to talk, as some social theorists do, of an "enlightened gratification" based on serving others when the "others" are nothing but machines.

This is, perhaps, the greatest danger of materialism. In making man a machine, we are, by analogy, making man into something that cannot be served unselfishly. It is impossible to be truly unselfish to your automobile, no matter how much you may treasure it. The tender care you give it, the scrupulous weekly washings, the regular check-ups, the oil changes, the lubrications, the concern over gasoline grades—you do all these because you want the car to serve *you*, not because you are serving the car.

Thus, by analogy, service to "nothing but" man makes sense only when the person you serve can do something for you in return. Truly unselfish behavior is possible only toward beings who are endowed with a conscious, spiritual element—an element which, by definition, makes them more than complex gadgetry. But such an element has no place in our scientific universe as it is generally conceived today.

Modern philosophers have done little to help us out of our dilemma. Until about 50 years ago, they boldly tackled the most vital problems mankind could face: questions about right and wrong, man's place in the universe, and man's nature and destiny. Their field was the queen of intellectual disciplines. But they long ago surrendered their intellectual primacy to the primacy of the scientists, succumbing to the philosophy of the "nothing but" man.

Modern science intimidates philosophers. They are afraid to challenge it or the conclusions they infer from its findings. Among philosophers, as among nearly all scholars, it has become almost taboo to ask questions about the human will, the mind, God, the meaning of life. These have been conveniently labeled "pseudo problems."

We urgently need philosophers to tell us how much we can validly conclude about ourselves from new findings in molecular biology and psychology. But the philosophers have cowered in their corner like kicked dogs. Their "philosophy" consists either of word games—designed, they tell us, to clarify issues—or bland credos such as existentialism, which has been dubbed "boredom brought to the level of a passion."

Yet we know virtually nothing about the relationship between the human brain and conscious thought. Theories of how the brain functions have had to be revised more often than our ideas about any other organ in the human body. Each time we come up with a new theory, an unexpected fact pops up to contradict it. There was a time when scientists thought each section of the brain had a special job—one section for sight, another for speech, and so on. This was replaced by the idea of "mass action" in the brain when it was discovered that removal of the speech section, for example, did not always mean a total loss of the ability to talk.

In 1966 a 47-year-old Omaha man further confounded matters by getting along quite nicely with only half a brain. The victim, a patient of Dr. Charles W. Burkland at the Omaha Veterans Administration Hospital, survived the removal of the left half of his brain in an operation in December, 1965. Eight months later he was surprising the medical profession by performing functions few would have thought possible under the circumstances.

Bit by bit, the patient somehow was able to regain abilities which, had they been stored entirely in the left side of the brain, would have been irrevocably lost. From uttering only a few words immediately after surgery, he soon progressed to the point where he could give clear renditions of "America," "Home On the Range," and assorted hymns.

Within less than a year he was able to pass simple mathematical tests, his color perception was almost perfect, he could write his name and the names of family members, and was slowly regaining much of the control over his right side.

The left side of this man's brain had been the "dominant" side—theoretically, at least, it had been the side that "carried the load" of the man's movements, perception, and speech. Theoretically, he should have lost the ability to speak and write. He did not.

In itself, of course, this cannot be used as an argument against materialism—after all, the man did have half his brain left. As was observed by Omaha neuro-psychologist Dr. Aaron Smith, the man's subsequent accomplishments suggest that "both hemispheres play a part in all mental functions, differing only in the extent to which they contribute." But it does show the folly of leaping to conclusions about similarities between the brain, thought, and computers—our understanding is so slight that we don't even know what can be accomplished by half a brain functioning alone!

A stronger challenge to materialism comes from Dr. Wilder Penfield, a pioneer in brain exploration. In a 1961 address at the University of California's San Francisco Medical Center, Dr. Penfield hinted at the existence of a non-material, spiritual agency that makes choices and, in effect, "wills" certain brain functions. This was strong stuff to hand out to a group of rigorous, scientifically oriented physiologists, psychologists, and other doctors.

Dr. Penfield told how, in his work in McGill University in Montreal, he was able to paralyze patients' speech by sending an electric current through the brain at a certain point. The patient doesn't know he can't speak until he tries, Dr. Penfield said. "Then," Dr. Penfield said, "he discovers to his astonishment that he cannot capture words."

Dr. Penfield continued: "One patient was being shown a picture of a butterfly when the electrode was applied to the speech area. He was silent. After the electrode was withdrawn he was asked about this. He explained that he could not get the word 'butterfly' so he tried for 'moth.' But that word, too, would not come."

In other words, Dr. Penfield went on, a person is able somehow to retain a concept—in this case, the concept of butterfly and moth—even when the words won't come. The patient was able to present these concepts, one after another, to the speech mechanism. Dr. Penfield called this "an example of the operation of the mind"—radical words, in an age which for the most part does not accept the existence of "mind" as a *distinct* entity at all.

The brain has many mechanisms, Dr. Penfield said, which "work for the purposes of the mind when automatically called upon." These mechanisms, he noted, comprise part, at least, of the physiological basis of the mind. Then he posed this profoundly un-materialistic question:

"But what agency is it that calls upon these mechanisms, choosing one rather than another? Is it another mechanism, or is there in the mind something of different essence?"

Continued Dr. Penfield:

> Surely we are nearer the goal [of complete understanding] than Aristotle was, and yet, after all this progress, when we try to see the actual link between the patterning of electrical impulses and a change in the mind of man, we are still in the dark —as much in the dark as Aristotle when he asked so long ago, "How is the mind attached to the body?"
>
> In conclusion, it must be said that there is as yet no scientific proof that the brain can control the mind or fully explain the mind. The assumptions of materialism have never been substantiated. Science throws no light on the nature of the spirit of man or God.

One gets the feeling that, in denying spiritual and/or religious realities, we are starving some deeper part of ourselves. We are, in effect, denying sustenance to a portion of ourselves for which nature has decreed regular feedings from some spiritual trough. In thus drying up the spiritual wellsprings of our being, we are developing a mood of despair, cynicism, and hopelessness.

Many people deny that man has spiritual needs. They point proudly to the fact that their own children are happy and well-adjusted even though they have been given no beliefs which, by any meaningful definition, could be called "spiritual." This is exactly like saying, "My children never ate an orange in their lives and they are as healthy as children who eat oranges regularly; therefore oranges are useless." It's like saying, "My father smoked heavily all his life and lived to be as old as most people who don't smoke; therefore smoking is harmless." Or like saying, "I am 40 pounds overweight, am 75 years old, and have never had a heart attack; therefore there's no connection between being overweight and having heart attacks."

In short, we can not be sure exactly what happens when an entire nation loses its religious and spiritual groundwork. Simply because a few individuals can be happy and healthy with no spiritual convictions doesn't mean that large populations can. Yet this is precisely the argument advanced by materialists, including some thoughtful scientists.

These same scientists would be extremely cautious about recommending a nationwide change of diet. They would be extremely careful not to over-generalize about the possible effects of small amounts of sulfur dioxide on the health of 500,000 city dwellers. They would never recommend that an experimental drug be put on the market after only two or three clinical trials. Yet they have no hesitancy at all in proposing wholesale changes in a nation's spiritual habits.

On the other hand, there are a few scholars who do believe that spiritual beliefs are important to a nation's social health and stability. There are a few people who think that much of our neuroticism and mental illness can be traced, at least in part, to a breakdown in our spiritual convictions. Listen to what the late Carl Gustav Jung, the pioneer psychoanalyst, had to say in *The Undiscovered Self:*

> Just as man, as a social being, cannot in the long run exist without a tie to the community, so the individual will never find the real justification for his existence, and his own spiritual and moral autonomy, anywhere except in an extramundane principle capable of relativizing the overpowering influence of external factors. The individual who is not anchored in God can offer no resistance on his own resources to the physical and moral blandishments of the world. For this he needs the evidence of inner, transcendent experience which alone can protect him from the otherwise inevitable submersion in the mass. Merely intellectual or even moral insight into the stultification and moral irresponsibility of the mass man is a negative recognition only and amounts to not much more than a wavering on the road to the atomization of the

individual. It lacks the driving force of religious conviction, since it is merely rational.

Our need for some sort of spiritual conviction must grow as the population grows. The more of us there are, the more we will need a belief in our own spiritual worth. With large numbers of people, the value of human life will decline unless we have more than materialism to maintain it. It's simple economics: scarcity determines value, and human life promises to be anything but scarce.

As the masses mount, we will find ourselves resentful, even contemptuous, of our fellow beings unless we believe in their ultimate spiritual worth. If we continue to view man as "nothing but," our contempt for human life will increase as men become more common. Reverence for life is not enhanced by long lines of human ants. Teeming masses of humanity do not stir respect for human dignity. Swarming crowds do not evoke concern for the rights of individuals. The more of us there are, the less we will think of each other—at least collectively—unless we firm up man's best convictions.

And otherwise the lower will be our opinion of ourselves, too. We could become, more than ever, mere numbers. More than ever, we might be like cogs in a vast machine—and not very important cogs at that. Crime could increase sharply—it might be each man for himself. The suicide rate could escalate. Our mental hospitals might be even more crowded than at present.

If all this takes place, we will become less and less mindful of the dangers of atomic war. We will be less and less horrified by the idea of large-scale killing. Mass murder may even seem attractive to some. Civilization itself will have lost much of its value. If someone pushes the fateful atomic button—well, so what?

What can save us? More than anything, we need a new self-consciousness. We need a re-examination of the assumptions on which materialism is based. We need to find out whether there is anything about mankind that does not fit the materialistic picture. We need a careful study of everything in man's experience which suggests that materialism may be in error.

I did not always think, as I do now, that an objective examination of man by scientific means could tell us more than the bare facts of our physical nature. I began life believing pretty much what all children believe—that God had a human form, wore a black robe much like our minister's, and that death did not exist. When I did learn that people die, I tried to picture what it must be like—black, black, black forever and ever and ever. No one tried to tell me there was a heaven or hell—no one but a neighbor whom I did not believe.

My father instilled a deep respect for science in me at an early age. Yet I did not then see it as a means to more than a surface understanding of man. What I did not understand then was that the methods of science can be used to find out whether or not seemingly unexplained, and psychical, events really occur. Science may never be able to tell how such things happen. But the question of whether they do or do not occur can be investigated by accepted scientific procedures.

To many scientists, the very idea that people can perceive each other's thoughts or see the future seems too wild to deserve serious consideration. And this may be partly why we're still laboring under a "nothing but" philosophy of man. The very unconventionality of the things that dispute the "nothing but" view is enough to threaten a scientist's reputation if he tries to study them seriously.

The events, the research, and the conclusions I am going to set forth in these pages may seem absurd to many, largely because they don't fit existing theories. But I intend to report them as faithfully as I would news of any kind.

I don't believe I am one to be easily "taken in" by fraudulent practices. The usual seances are distasteful to me. A session with a typical medium leaves me with such a sour taste that I'm sometimes inclined, after such a sitting, to want to forget the whole field of psychical research and turn to more solid pursuits. But I can't blindly dismiss some of those things which I know are not fraudulent, outlandish as some of them seem. Especially not when they are so important to our understanding of our nature and to the kind of world we make for ourselves.

* III *

The Search for a Repeatable Experiment

We were both 12 years old. For more than a month we had been practicing. He was to play the part of a turbaned Oriental swami. I was to be the classical, high-hatted magician. His attic had been converted into a theater. Our parents, friends, and relatives had been invited.

With the big show only 24 hours away, we decided to take a break. We sat down on the floor at the far side of the room. He casually picked up four ordinary playing cards, each from a different deck. Each card had a different picture on its back. One of us suggested we try some real mind reading.

I had never heard the term "extrasensory perception." I certainly had no notion of the laboratory ESP experiments that had been under way at various universities here and abroad for many years. My friend and I were acting on a simple whim, then, when he stared poker-faced at one of the cards and asked me to guess what was on it.

I concentrated for perhaps a quarter of a minute. I knew the card showed either a train, a plane, a ship, or an American flag. But I had no way of knowing which of the four he was looking at.

No "normal" way, that is.

I don't recall, now, which card I chose. But I do remember my guess was correct.

So was the next. And the next. And the next.

Then we switched. He did the guessing while I looked at the cards.

He, too, was right every time.

I don't remember how long we kept it up. But I do remember that, amazingly, we seemed never to miss. I can almost feel again the thrill that grew and grew each time I made a guess and found that it, too, was right.

High excitement claimed us. It seemed we had stumbled onto something big. Maybe we were really mind readers! What a show we'd put on the next night!

And then it stopped. The moment we got excited, the moment we became a bit anxious over our performance, we lost our seeming ability.

I didn't know it then, but this was consistent with findings being made in the widely publicized experiments by Dr. Joseph B. Rhine, then at Duke University. Using cards, Rhine was finding that people perform best in ESP tests when they're in a playfully competitive spirit. Their scores drop to chance levels, he found, when they become either over-eager or bored.

Our off-hand experiment was far from proof of anything like ESP, of course. Ironically, many of the chief weaknesses in our little game also showed up in the early work of Dr. Rhine. For one thing, we could see each other's faces—and, of course, we could have picked up telltale changes in each other's expressions which, without our realizing it, could have influenced our guesses. We could also see the cards—and although the pictures were carefully concealed from whoever was doing the guessing, slight imperfections in the cards themselves might have transmitted subtle clues which we could have picked up without realizing it. Finally, our cards were shuffled by hand instead of being systematically randomized.

Dr. Rhine and his staff took pains to avoid these deficiencies in their experiments.

For those who are unfamiliar with Dr. Rhine's experiments, a word of explanation may be in order here, since much significant work is being carried out today with variations on the techniques he helped pioneer.

Dr. Rhine used a deck of 25 cards. Each card showed either a square, a circle, a star, a cross, or parallel wavy lines. There were five of each kind of card in the deck. Thus, if you guessed five correctly out of 25, you'd be doing only as well as could be expected by chance. But if you scored above that, and if you kept it up over a period of hundreds of trials, something other than chance would be needed to explain your success. Dr. Rhine coined the term "extrasensory perception" to describe this. Sophisticated statistical procedures were used to determine how far above chance a given subject was performing.

Using this method, Dr. Rhine and his colleagues found

that some people were able to "guess," time and time again, which card another person was looking at. This, he concluded, was evidence for telepathy or thought transference—direct mind-to-mind contact.

He also found that some people could "guess," to a statistically significant degree, the order of the cards in the deck even when no one else was looking at them—evidence, he concluded, that the mind is somehow able to perceive objects and events at a distance, with no mediation by the physical senses. This ability is known as clairvoyance.

Thirdly, Dr. Rhine found some subjects could apparently predict, with better-than-chance accuracy, the order the cards would be in after they were shuffled. Evidence, he concluded, for precognition—the ability to foretell the future.

Dr. Rhine's work proceeded from a scattering of other research carried out by independent investigators, including such distinguished thinkers as psychologist William James, philosopher Henri Bergson, psychiatrist-neurologist Pierre Janet, and even—on a speculative level—that esteemed old Greek, Democritus. Controlled experiments, not unlike Dr. Rhine's own card-guessing tests, had been made in England, Sweden, Poland, France, and Germany. Telepathy experiments were carried out by, among others, French engineer René Warcollier; Dr. Carl Bruck, a German physician; and American novelist Upton Sinclair.

In Sinclair's work, his wife successfully duplicated drawings on which another person, often miles away, was concentrating. Albert Einstein was among those who were impressed with Sinclair's work—so much so that he felt Sinclair's book, *Mental Radio*, should be given a scientific hearing.

In addition to these attempts at controlled verification, volumes of spontaneous case material had been investigated and compiled by various psychical research societies.

Few of these projects, however, received as much attention as Dr. Rhine's. Psychologists criticized his early work because the experimenter and subject were in plain sight of each other and the order of the cards was not scientifically randomized.

Although Dr. Rhine corrected these defects in subsequent experiments, his work is still occasionally criticized on these same grounds—usually in informal discussions between professors and their students.

There is, however, another difficulty with Dr. Rhine's

The Search for a Repeatable Experiment

work. And not only with his. It has plagued virtually every experimenter who ever tried to study ESP under laboratory conditions. It remains, today, the biggest stumbling block to general scientific acceptance of ESP.

It is the problem of repeatability. Any experimental effect, if it is to be taken seriously by a majority of scientists, must be capable of being repeated at will by many scientists in many laboratories. This has not been true of ESP experiments.

To get positive results in a typical ESP experiment, you first have to find a good subject. Good subjects are scarce. After you've located one—assuming you're fortunate enough to do so—you must exercise special care to test him while he's obtaining high scores. Even the best subjects show a "decline effect"—that is, they do well for a time, and then their performance levels drop to chance.

These difficulties have invited charges that ESP investigators choose subjects who happen, by chance, to be having a "run of luck." The critics note that, if you look long enough and hard enough, running tests on enough people for a long enough time, you're bound to come across a few who—by chance alone—are getting an unexpectedly large number of hits. The so-called "decline effect," they charge, is simply the termination of the "run of luck."

Thus, the critics say, the ESP researchers load the dice in their favor, reporting their findings for only those subjects who happen to be getting above-average scores and only *while* they are getting those scores.

To meet this objection, many ESP researchers—known professionally as parapsychologists—are trying to bring repeatability into their experiments. This could be done very neatly if, in some way, parapsychologists could find a pattern to ESP performance which would enable them to state definitely that a certain type of person, in a certain mood, under specifiable conditions, would consistently produce results suggesting the operation of ESP.

Presumably, other scientists would then be able to duplicate the results by testing the same type of subject, in the same mood, under the same conditions.

To this end, many parapsychologists are hard at work looking for correlations between ESP performance and such things as personality test scores, physiological variables, intellectual abilities, sex and age differences, cultural and ethnic traits, and mood changes. The search has been extended to subjects' blood pressure, electrical brain impulses, respiration, and even perspiration.

Typical is the work being done by Martin Johnson, a psychologist at Sweden's Lund University, and B. K. Kanthamani of Andhra University in Waltair, India. These two researchers set themselves the task of finding out whether people who are prone to anxiety are either consistently good—or poor—ESP subjects.

We have good reasons for thinking there might be some connection between anxiety and ESP performance. Anxiety may block material coming from deep in the unconscious, for example, thereby inhibiting the operation of ESP—that is, assuming ESP is an unconscious process.

Presumably, if one could identify good ESP subjects by means of an anxiety test, a significant step would have been taken toward repeatability.

Johnson and Kanthamani gave a recently developed anxiety test and a standard ESP card test to 27 persons—16 in an initial experiment and 11 more in a follow-up. The anxiety test was designed, say the researchers, to probe for the presence of anxiety at an unconscious level. Unlike many of our straightforward question-and-answer anxiety measures, it involved the use of pictures flashed on a screen. Each picture showed a centrally placed person—a "hero," if you will—with whom the subject supposedly identified himself. This central figure was always in a dangerous, threatening, or compromising situation. Initially the pictures were flashed so fast the subjects could hardly see them. In succeeding showings the rate was slowed until, bit by bit, the subjects were able to give increasingly accurate descriptions of the pictures.

The subjects' distortions of the pictures—what they thought they saw, but saw incorrectly—gave clues to key aspects of their personalities. The aspect being specifically measured was, of course, anxiety.

Less anxious subjects, the scientists found, got more consistently high ESP scores. Conversely, the more anxious a subject was, the lower his ESP performance. The very anxious subjects got consistently negative ESP scores—their performance was below what would have been expected by chance.

Johnson and Kanthamani both say more research is needed using this test, known as the Defense Mechanism Test (DMT), before we have really conclusive evidence of an anxiety-ESP relationship. Other anxiety tests—the simpler, more straightforward kind—haven't correlated consistently with ESP. But these other tests, as Johnson and

Kanthamani note, measure anxiety on a purely conscious level.

The scientists suggest that we think of ESP messages as being channeled from the unconscious into awareness. These messages tend to be blocked, they note, in persons who have highly developed defenses against material from their unconscious. This would include anxiety-prone individuals. Hence the importance, according to Johnson and Kanthamani, of using a test that measures anxiety on an unconscious level.

Kanthamani found another apparent relationship between ESP and personality. While watching a casual ESP game between pairs of high school students, she noticed that one tended to get above-average scores consistently while, just as consistently, his competitor got low scores.

The relationship held up, she found, whenever any two students tried to out-ESP each other. One member of each pair would repeatedly tend to score positively and the other negatively.

So she carried out a formal study of 53 students. Not only did this unexplained relationship persist, but she found that she was able to tell, in advance, which member of each pair would get the high scores and which the low! Through technical statistical procedures she found that the odds against these results being due to chance were 100 to one.

Her research may be another small but significant step toward predicting ESP performance—and hence toward repeatability. There's only one problem: her ability to predict which member of the pair would score positively was based on her own personal clinical interpretations. These interpretations, she says, were drawn from her observations of the children's behavior, their reactions to each other's success and failure, and her knowledge of their performance during their first few trials.

Her results, then, are promising, and suggest a possible avenue for further research—a study to find out what, exactly, her clinical impressions were based on. If an objective scale could be derived from these impressions, perhaps other experimenters could also predict which student would get the higher scores. Accomplish this, and you have repeatability.

In another experiment, Dallas E. Buzby, of the parapsychology laboratory at St. Joseph's College in Philadelphia, found he was able to predict people's ESP performance by the amount of interest they expressed in ESP and by their scores on a simple sense perception test known as the Embedded Figures Test.

People who were seriously interested in ESP, Buzby found, showed more evidence of ESP ability than those who were only casually interested—a result that's consistent with findings of other parapsychologists.

But Buzby went further. The Embedded Figures Test, he found, could be used to separate people whose ESP scores hovered consistently around chance levels from those whose ESP performance varied more widely from low to high.

Actually, the Embedded Figures Test merely measures a person's ability to identify a simple design embedded in a complex one. Supposedly, if you can pick out the simple design in less time than average you are analytically minded; if it takes longer than average you are globally minded—or so say the test's designers.

Among the 78 people Buzby tested, the analytically minded—in this case, those who could pick out the simple design in less than 35 seconds—got ESP scores that clustered around chance levels even more than would be expected by chance! In other words, the odds are that on any ESP test your scores won't be right at chance levels all the time—they'll fluctuate a bit. If they are at or near chance consistently, something other than chance is operating. This, Buzby found, was true of the analytical perceivers.

The global perceivers—those who took 37 seconds or more to pick out the design—tended to fluctuate away from the chance average more than could be expected by chance alone.

Buzby found that he could predict ESP performance even more accurately by combining the Embedded Figures Test with an interest questionnaire. If you are highly interested in ESP and are also a "global perceiver," you'll have consistently wider variations in your ESP scores than you will if you are analytic and are only casually interested in ESP, Buzby discovered.

Critics may argue that the Embedded Figures Test itself is not among the most scientifically validated psychological measuring devices. But even if it doesn't measure exactly the things its originators say it does, the fact remains that the test apparently can predict ESP performance—which means, of course, that it may be useful in approaching ESP repeatability.

Many of the most interesting and intensive studies in

this field have been carried out by Dr. Gertrude Schmeidler, of the psychology department at The City College of the City University of New York. Dr. Schmeidler ranks among the foremost searchers for an ESP-personality link. In cooperation with Dr. Robert A. McConnell, a biophysicist at the University of Pittsburgh, she has made a pioneering search for a relationship between ESP and personality patterns as revealed in a wide variety of personality tests, including the well-known Rorschach ink blot test. Rorschach protocols, she found, could be used with some consistency to predict ESP performance.

Promising attempts to find consistent ESP differences between boys and girls have been made by John Freeman, of the Institute for Parapsychology in Durham, N.C. Freeman has also carried out experiments which suggest that *conscious* anxiety may actually facilitate ESP performance —in contrast to the effects of unconscious anxiety, as measured by Johnson and Kanthamani. Other attempts to correlate conscious anxiety with ESP performance have given conflicting results, however.

Dr. Carroll B. Nash, a biologist at St. Joseph's College, is another leading seeker for an ESP-personality relationship. In one study, Dr. Nash found that, of ten traits measured by a standard personality questionnaire, only one— "general activity"—showed a strong positive correlation with ESP. The other traits measured by the test, called the Guilford-Zimmerman Temperament Survey, are sociability, emotional stability, objectivity, friendliness, capacity for personal relations, restraint, ascendance, thoughtfulness, and masculinity.

"General activity" includes such qualities as drive, energy, action, vitality, speed, and efficiency. Suggests Dr. Nash: "A propensity to general activity may increase the energetic action applied in the ESP test and thus increase the ESP score. Also the ESP score may be higher in the person with high general activity because he is less inhibited in his responses, including those to ESP."

And so the search for a repeatable experiment goes on. The experiments cited here are only a few of the many studies being pursued in an effort to find an experiment which will convince even the most stubborn skeptics.

A number of psychic investigators feel the laboratory parapsychologists have become obsessed with card guessing and statistics. The most significant, interesting material, they feel, occurs spontaneously—like the mother who dreams her Air Force son is being shot down in flames at the exact moment it happens. They feel that studying psychic phenomena is like studying a tornado or an earthquake, neither of which is a repeatable event.

This, then, raises the question: just how important is repeatability? If an effect cannot be replicated at will, does this exclude it forever from the halls of science? Are scientists justified in ignoring it on these grounds?

The answer, of course, is No. Scientists do not ignore astronomical and meteorological events simply because men can't conjure them up in a beaker. What scientist would deny the reality of sun spots simply because they have never been duplicated inside his own university quarters?

This has led a few, such as the British biologist Sir Alister Hardy, to conclude that parapsychology should be classified as a natural science. Studying psychical phenomena would then be a little like studying the behavior of gorillas in their native environment. You can't very well catch your subject matter and bring it home, and still be observing the phenomena you set out to investigate. You have to take it where you find it, study it when you see it, and leave it where it is.

Why, then, don't orthodox scientists accept parapsychology on these terms?

For two reasons, I think. First, the material studied by parapsychologists has long been associated with mythology, superstition, and fraud. Scientists, on the whole, are justifiably proud of science's record in debunking superstition. Many of them view science as a sort of positive replacement of mythology in man's search for the truth. Genuine progress, if such exists, is seen as the growth of scientific understanding in the face of folklore and hokum.

It follows, then, that few scientists want their names linked with anything that smells like superstition. A scientist who shows a serious interest in occultism loses status in his profession. His fellow scientists view him with a mixture of scorn and pity.

Students, who frequently express curiosity about ESP, soon find their interest squelched by their professors. A Pittsburgh biologist once told me that, when any of his students showed a persistent interest in ESP, his answer was:

"Perhaps the field of science is too confining for you—perhaps you should consider a different career."

The biologist was proud of that reply. He felt it was fair and impartial. When he asked me my opinion, I suggested it might be better to warn the students that ESP researchers generally get a rough time from their colleagues, and then add that perhaps science is too confined—perhaps science could use more exposure to unorthodox concepts.

In another case, a University of Pittsburgh professor was giving his students a lecture on the work of Prof. William McDougall, who is known, among other things, for his controversial studies of the inheritance of acquired characteristics. Commented the professor:

"Among the crackpot things McDougall was interested in was ESP!"

Imagine many of those students taking a serious interest in parapsychology later! Without any attempt at a careful presentation of the facts, that professor probably planted a permanent seed of unbelief in each of those student minds. How many of them will ever, throughout their entire careers, bother to look seriously at evidence touching on ESP?

The other reason orthodox scientists don't accord parapsychology the status of a natural science is that there is no good, mathematically based theory to account for the data.

Most scientists are motivated by, among other things, a need for understanding. They like order. To many of them, the picture that science is revealing of the universe is as esthetically pleasing as great classical music. It all fits together in a huge symphonic relationship. The motions of the largest, most far-flung stars obey the same basic laws as the RNA molecules that help decide the shape of living things. Even the unpredictable world of the sub-atom can be fitted in mathematically.

When something as outrageous as ESP comes along, then, most scientists reject it intuitively, emotionally, esthetically, and intellectually. If there is no theory to explain it, they feel it cannot be true, no matter how persuasive the data. Lacking a theory, they feel justified in rejecting it on the grounds of unrepeatability. Without a model, it can't qualify even as a natural science. Students are urged not to "waste time" on it.

Some scientists apparently feel threatened by ESP research. They react angrily to it. When *Science*, the journal of the American Association for the Advancement of Science, published a study of apparent telepathy between identical twins in its Oct. 15, 1965 issue, a storm of protest ensued. The report was heavily criticized on technical grounds. Many of the criticisms were justified. But the emotional content of many so-called "scientific" responses suggested that the scientists disapproved of the subject matter itself. Harvard's Dr. Edwin G. Boring, who until his death was one of the leading scholars of the history of psychology, and a respected critic of parapsychology as well, wrote in a letter in the Dec. 3, 1965 issue of *Science:*

"The report of Duane and Behrendt [the Philadelphia scientists who did the experiment] . . . has so heated the mail to my usually quiet ivory tower that I now need insurance. One non-parascientist even asked: 'Ought I not to resign from the AAAS?' "

Dr. Boring went on to enumerate some entirely reasonable criticisms of the experiment, and then cited a case in which parapsychologists had once taken a serious interest and which later turned out to be a fraud. Presumably, the latter was an attempt to show that fraud could explain Duane and Behrendt's findings also.

Why *Science* chose to publish a report that didn't have nearly as much good, quantifiable data as other ESP experiments remains a mystery.

Some of our leading scientists seem perfectly willing to criticize experiments in ESP without taking the trouble to familiarize themselves with the data. Such was the case when Dr. B. F. Skinner, one of the nation's most respected psychologists, and author of the popular *Walden Two,* criticized a historic experiment which Britain's Dr. S. G. Soal and Mrs. Mollie Goldney carried out with an exceptional subject, Basil Shackleton. The Soal-Goldney experiment was discussed in the Spring, 1948 issue of *The American Scientist* by Yale's Dr. Evelyn Hutchinson in the column, "Marginalia." In its Summer issue, *The American Scientist* printed a critical letter from Dr. Skinner and a reply from Dr. Soal. It was clear that Dr. Skinner had never read Dr. Soal's report. For instance, Dr. Skinner criticized the use of hand-shuffled packs of ESP cards as targets. Dr. Soal noted, in his reply, that he had not used packs of cards in these particular experiments and that no shuffling procedures were involved in the research.

More recently, we have the case of Professor C.E.M. Hansel, whose book *ESP—A Scientific Evaluation* (Charles Scribner's Sons, 1966) earned him plaudits from noted psychologists, including an introduction by the aforementioned Dr. Boring. In essence, the book is an attempt to show that much so-called ESP can be explained more easily in terms of fraud than in terms of the ESP hypothesis. By implication, Hansel manages to malign the

The Search for a Repeatable Experiment

names and reputations of many parapsychologists whose experiments, he thinks, must have been fraudulently conducted because ESP is (to Professor Hansel) so improbable.

In a review of the book, Dr. Ian Stevenson, former chairman of the University of Virginia's psychiatry department, notes that Hansel makes no less than nine mistakes in a space of 22 lines in discussing just one experiment—mistakes about how the cards were handled, what results were obtained, and so on. "It is difficult to believe he [Hansel] has read any of the published reports of this work," Dr. Stevenson says.

What, then, can we say of the scientific status of parapsychology? We can state that, so far, we do not have the kind of evidence that can be demonstrated in a laboratory on demand. Nor do we have a mathematically precise theory that integrates the data into our total picture of the universe.

But we do have the kind of evidence obtained by explorers in remote regions who study rare, seldom-seen species of wildlife—creatures which, say, have consistently eluded captivity.

And the evidence itself, allowing for the problem of repeatability, is so compelling that we can not afford to ignore the subject. It is urgent, if we are to have a balanced and accurate understanding of ourselves, that people pay serious attention to parapsychology. No one who is informed on the subject can disregard it and still remain honest with himself.

* IV *

Telepathy and Clairvoyance

A young Pittsburgh housewife was seated in a café with her husband in the summer of 1965. "Suddenly," she later told me, "the room and everything around me seemed to disappear. In their place I saw a four-lane highway. It was night. Two motorcycles were coming around a curve. The first one skidded and crashed into a guard rail. The second crashed in half."

She mentioned the experience to her mother.

Two days later she learned a former school friend had been killed at precisely the moment of her vision. Accounts of the accident closely matched her description.

Parapsychologists say that an experience like that could be either telepathy—communication from the mind of one of the victims or a witness—or clairvoyance, a direct awareness of the accident itself. But coincidence, they insist, seems most unlikely in such cases.

Many such stories can, of course, be dismissed as fraudulent, coincidental, or simply bad observation. Others are very possibly due to slight—but entirely explainable—sensory cues. Cases that can be "explained away" fairly easily are responsible for a good deal of scientific skepticism about ESP—and justifiably so. Any explanation which can be understood in terms of known scientific concepts, and which still fits the facts, is preferable to one which would require a scientific revolution.

This is why organizations such as the British and American Societies for Psychical Research take such great pains to investigate cases of this nature, ruling out those that can be explained in normal terms and obtaining independent corroboration from as many witnesses as possible on any others.

The following account is backed up by precisely this kind of investigation. It was reported by the British Society for Psychical Research in its September, 1966 *Journal*.

An 85-year-old North Devon clergyman, identified as the Rev. A. B., sat down to brunch about 12:30 p.m. March 11, 1965. No one else was in the room or the house. Suddenly he heard a man's voice say clearly, "I am here." He recognized the voice as that of his nephew, Charles.

The voice, according to the clergyman, was "distinctly an external one." He turned abruptly to locate it. No one was near. Bewildered, he thought perhaps Charles was on the premises outside, about to pay his old uncle a visit. But no one rang the bell, and there was no one at the door.

The Rev. A. B. noted the experience in his diary.

Later that same day he read his nephew's obituary in the *Daily Telegraph*.

He described the experience to two visitors that day, both of whom were contacted by the Society for Psychical Research, and who gave written accounts that agreed with that given by the elderly clergyman.

Not content to leave any possible natural explanation unexplored, the Society asked the old gentleman if, perhaps, he had inadvertently heard the news of his nephew's death on the radio the night before. He assured them that not only hadn't he had his radio on, but that he was partially deaf so that he wouldn't have heard it inadvertently in any case.

Still probing for a natural cause, the Society questioned him on the possibility that he had unconsciously seen the death announcement in the newspaper as it lay on an easy chair a few feet away from him that noon. No, he said, he was blind in one eye and can't read a paper without a magnifying glass.

Was this a clairvoyant—or perhaps telepathic—perception which his unconscious mind had dramatized into a representation of his nephew's voice? We know that our minds do dramatize our fears and desires in our *dreams* along much these lines—so why not when we're awake, using ESP? This is known as sensory dramatization.

Another example of sensory dramatization was given me by a 71-year-old Buffalo woman. When she was 23, Florence C. Jones was frolicking with a group of friends in Cazenovia Creek near Buffalo, N.Y. Suddenly, to her terror, she found she could no longer touch bottom, having apparently drifted over a deep hole. She was unable to swim. A current sucked her under.

"I was sure I was going to drown," she said. She was below the surface, going down to what she feared would be a rather damp grave. Unexpectedly she heard the voice of her dead mother sternly command her to "put your arms over your head."

"It was the last thing I wanted to do," she recalled. Such a move, she said, would seem only to facilitate her downward progress.

Again she heard the firm voice ordering her to extend her arms upward. Against her common sense, she did as the "voice" directed. ("What did I have to lose? I was going to drown anyway.")

As it happened, however, this one movement probably saved her life. Her companions, in a valiant effort to save her, had linked arms and formed a human chain toward the spot where they had last seen the struggling young woman. They seized her upward-thrust hands and dragged her to safety.

"I had no idea how putting my arms up would help," she told me. "It seemed like that would only make me sink faster."

Clairvoyance? Telepathy? Or more?

In October, 1916, while my mother was attending Smith College, she injured her knee playing field hockey. The injury was exceedingly painful. Much of her leg had to be encased in a plaster cast.

Between the painful injury and the loneliness and frustration of being cut off from normal college living, she spent the next few weeks in utmost misery. Late one night, as she lay sleepless and acutely distressed, she yielded to an impulse to call her mother aloud—"Oh mother, mother!"

Several days later a letter from her mother told her that, on the night in question, her mother had awakened hearing the girl calling her.

In this case the older woman had already been informed of the injury and, presumably, suspected her daughter was unhappy. Still, the fact that she heard her daughter calling her at about the time the girl actually was calling seems more than coincidence.

In each of these three cases, the percipient's mind apparently dramatized a message from deep in his or her unconscious mind—perhaps a telepathic or clairvoyant message

Telepathy and Clairvoyance

—presenting it as the voice of someone he or she loved. But our minds can play even neater tricks than that.

Mrs. Patricia Packard, a Pittsburgh housewife, sat exhausted in her living room after a particularly trying day late one afternoon. Her 17-year-old son, Andy, was upstairs in his bedroom. Dinner was almost ready.

"I was too exhausted to move," Mrs. Packard told me. "I knew it was time to tell Andy to come down to dinner, but at that point I didn't feel up to climbing the stairs."

With Andy still in mind, she apparently drifted off into a semi-conscious state between sleeping and waking. Minutes later she roused herself, to find that Andy was headed downstairs.

When the boy saw his mother, his eyes fairly popped out of his head, his mother said later. "How," he asked her, "did you manage to get down here so fast?"

"What do you mean?" Mrs. Packard asked.

"Why, you were upstairs just a second ago telling me to come down," he stated.

He insisted, over his mother's objections, that she had actually come into his room and told him to get ready for dinner.

Had Mrs. Packard, in her slumber-like state, somehow projected a telepathic image of herself upstairs? Had Andy himself perceived, extrasensorily, his mother's wish, and dramatized it visually?

Or had Mrs. Packard, as our more skeptical friends would insist, actually walked upstairs while she was half asleep, told Andy to come down to dinner, and then forgotten about it? But if that's how it happened, we must ask with Andy: how did she get downstairs so fast?

Mrs. Packard has had other telepathic—or, perhaps, clairvoyant—experiences. "They bother me," she told me one day on the phone. Seized by a whim, I held out four fingers behind my back (why I held them behind my back I'll never know, since, being on the other end of the telephone line, she couldn't have seen them in any case).

"How many fingers am I holding out?" I asked.

"Four," she replied, without a moment's hesitation.

"Now how many?" I asked, holding out three.

She paused. She wasn't sure. She said she kept getting the number "one," but added she was certain that was wrong.

The interesting thing about that brief little guessing game was Mrs. Packard's unhesitating certainty when she was right, and her equal certainty that she was wrong when she missed.

Being of too impatient a nature to conduct extensive guessing-type ESP experiments, I did not pursue the matter further.

Like many people, I have never been temperamentally suited to long hours of guessing cards—or conducting experiments in which other people try to guess them. But scientific experiments in telepathy and clairvoyance have now broadened far beyond the standard card-guessing tests. While the latter are still basic to a great deal of ESP research, imaginative investigators have come up with fascinating ways of testing telepathy in dreams, under strong emotions, and in novel situations.

A scientist in Tasmania, for instance, substituted music for cards. H.H.J. Keil, a lecturer in psychology at the University of Tasmania, got significant results from having his subjects guess which of five musical numbers was being played in a room some distance away.

And at the University of Witwatersrand in Johannesburg, South Africa, Professor A.E.H. Bleksley, head of the Department of Applied Mathematics, tried using time itself as the target. His subject, a Mr. W. van Vuurde of Cape Town, was one of those fortunate people who can wake up at any previously decided-on time without an alarm clock. For some reason, van Vuurde figured maybe ESP was involved in the process. He set out to test this by having himself wake up at all kinds of unholy hours during the night, chosen at random without his conscious knowledge.

His results, obtained over a period of months, looked so promising that he sent a glowing report to Professor Bleksley.

Bleksley was impressed. "To me, the results appeared, quite frankly, so remarkable as to border on the fantastic," he said, adding that he saw this as an excellent chance to test a confident and enthusiastic subject by a method the subject himself had chosen and enjoyed.

Van Vuurde had been trying to wake up at the time designated on an old, broken, stopped alarm clock whose hands had been set at a randomly chosen position the night before. At Bleksley's request he sent the clock to Johannesburg, where the mathematics professor planned to set it himself each day. Then van Vuurde, more than 900 miles distant, would try waking up at the target time.

The first experiment was less than successful. The results were suggestive, but not statistically significant.

In a series of experiments carried out *after* the two men had met, however, van Vuurde's internal "ESP clock" seemed to be functioning most successfully. Some of his scores couldn't be explained by mere chance. Throughout, the sleeping van Vuurde showed a remarkable tendency to be either exactly right, or way off—he either woke precisely at the minute designated, or he didn't even come close. It looked as if, when he missed, it was because something other than his own ESP was waking him up—we all sometimes wake up for no obvious reason during the night, and van Vuurde was apparently no exception.

Does your ESP work better, then, when you're asleep than when you're awake? It sometimes seems so, the way ESP apparently keeps showing up in dreams. Two scientists at Brooklyn's Maimonides Hospital have set up a special "dream laboratory" in which, night after night, they have tested sleeping people's abilities to dream about prints of famous paintings being viewed by another person—the "sender"—in another room.

Dr. Montague Ullman and Dr. Stanley Krippner, of the Maimonides Psychiatry Department, worked out an ingenious technique whereby each of their sleeping subjects' dreams could be compared with the very picture—never seen by the subjects—that the sender had been looking at when the dream was taking place. The scientists used two well-established procedures which show, almost infallibly, exactly when another person is dreaming: they watched the movements of their slumbering subjects' eyes beneath their closed lids, and they observed the graph-like tracings on a device that recorded the slight electrical impulses given off by their brains. When a sleeper's eyeballs move rhythmically from side to side (or up and down), it's a good indication that he's having a dream—apparently one follows the action in one's dreams with the eyes, much as if one were watching a movie. At the same time, if one is hooked up to an electroencephalograph—an instrument that picks up the slight electric impulses emitted by the brain through wires painlessly fastened to one's head—a trained observer can tell that one is dreaming by the changes in the electrical brain wave patterns recorded by the encephalograph.

It was a simple matter, then, for Ullman and Krippner to note when each of their subjects was dreaming, to wake him up as soon as the dream was over, to have him tell his dream to a tape recorder, and then to let him go back to sleep.

One researcher recorded the time of the dream, while another made a note of which picture the sender was looking at in the other room. (Each subject had been told to "try" to dream about whatever picture was being observed by the sender.)

Now, here's the key to the whole experiment: the scientists did not know until later which picture corresponded to which dream! They gave the pictures and written descriptions of the dreams to three judges they had chosen, instructing these to try to match the pictures and dreams. The judges were asked to indicate which of the 12 pictures they thought most resembled a given dream, which was the second most similar, which was third, fourth, and so on. Then the scientists had the subjects themselves try to match their own dreams with the appropriate pictures, again ranking them in order of similarity. In a third phase of the experiment, the subjects' free associations to their dreams were written down and the judges tried to match the pictures with these associations.

The scientists took strict precautions to prevent any possible communication between the sender and the sleeping subject. All conversation between experimenter and subject on dream night was mediated through an intercom system and recorded on tape as a guard against references to the target pictures.

After a detailed statistical analysis of the results, Drs. Ullman and Krippner concluded that the subjects themselves were able to match the pictures with their dreams more accurately than could have been expected by chance. The judges, they found, were able to match pictures and dreams also, provided they had the subjects' verbal associations to the dreams to refer to. Surprisingly, however, the judges couldn't match the dreams and pictures to any really significant extent without those associations!

But this was only the beginning. Although the results with the initial 12 subjects—seven men and five women—looked promising, they were hardly what you'd call spectacular. In succeeding experiments, Dr. Ullman and Dr. Krippner selected high-scoring subjects and tested them separately. In one seven-night series of dreams, a single subject—identified beforehand as one who was likely to dream successfully about "target" material—turned in a performance that should stir the interest of even the most cautious scientists. The three judges, each one matching

Telepathy and Clairvoyance

the dreams and pictures independently of the others, were able to tell which picture the subject had been dreaming about to an extent that ruled out chance by a factor of 100 to one.

And in later experiments the scientists obtained still more striking results. One subject, a psychologist, got 19 direct or near "hits" in 20 of his transmitted dreams!

What are we to make of this? It's one thing to collect case histories of seemingly telepathic dreams people say they had, and it's quite another to tell a person to dream about a certain thing without telling him what that thing is.

The Maimonides dreams weren't as detailed as many of the apparently telepathic dreams outside the laboratory have been. Here, for instance, are a few excerpts from dreams one subject at Maimonides had over a period of seven nights:

"There was one scene of an ocean.... It had a strange beauty about it and a strange formation.... Fishing boats ... I had some sort of a brief dream about an M.D.... The picture ... that I'm thinking of now is the doctor sitting beside a child that is ill.... It's one of those classical ones.... It's called 'The physician.' "

The target picture was the famous painting, "The Sacrament of the Last Supper," showing Christ at the table with his disciples and, in the background, a body of water and a small boat.

Nothing too amazing there, you say—a few references to water and boats and a doctor, so what?

But wait. Listen to what the subject had to say when he was questioned about this dream after he woke—listen to his verbal associations:

"The fisherman dream makes me think of the Mediterranean area, perhaps even some sort of Biblical time. Right now my associations are of the fish and the loaf, or even the feeding of the multitudes.... Once again I think of Christmas."

Now, suppose you were one of the judges and had to decide whether this dream was most like "The Sacrament of the Last Supper," Van Gogh's "The Starry Night," "Animals" by Rufino Tamayo, or "Football Players" by Henri Rousseau—all of which were used in the experiments. Even if you had never seen the pictures themselves —a condition that wasn't imposed on the judges—which would you choose? Just from the titles alone? Which picture has a Biblical Mediterranean setting, is associated with such New Testament stories as the loaves and fishes, and might suggest Christmas?

No, it's hardly like the detailed dreams of tragic fires and accidents that fill the files of psychical researchers. It may be even more significant than those. It's an actual laboratory demonstration of people dreaming, crudely and incompletely perhaps, about specified targets that they could not have perceived through their normal senses. Distorted dreams, yes, but accurate enough so that other people could, in some cases at least, match them with the targets.

And these were dreams about mere pictures—pictures being looked at by someone who wasn't related to the subjects. What happens when you get favorable conditions for telepathy between two closely related persons—between, say, a husband and wife?

Dr. Marie Coleman Nelson, a Smithtown, N.Y. psychoanalyst and former managing editor of the scholarly *Psychoanalytic Review*, tells of an estranged man and wife who had a remarkably similar series of parallel dreams May 11, 12, 13, 14, and 16, in 1948. After ten years of marriage, their feelings toward each other had become so negative that they no longer conversed, had sexual relations, or visited friends together. The only activity they shared was breakfast, and the only interest they shared was psychoanalysis, which both were undergoing.

Their interest in psychoanalysis led them to relate their dreams to each other over their morning coffee—this being one of the few topics that didn't trigger quarrels.

Their dreams were so similar, on May 11 and 12, that the wife—whom Dr. Nelson calls Stella—decided to write down her dreams each morning before seeing her husband. She also kept a written record of the dreams of her husband, whom Dr. Nelson calls Kurt.

The dreams, as reported by Dr. Nelson in the October, 1964 issue of the *Journal of the American Society for Psychical Research:*

Date	Kurt's Dreams	Stella's Dreams
5/11/48	A girl is proudly displaying a long tail—hair or fur.	A girl is proudly displaying her long, straight hair.
5/12/48	I try to phone Stella and do not succeed. I am annoyed with her.	I am having trouble with the telephone. It either rings, and when I answer nobody is there, or I expect it to ring and it does not.

Telepathy and Clairvoyance

5/13/48	I am in an elevator. The operator wants it to go down and I try to take the controls and make it go up.	I am in an elevator. On a floor below I see through the grillwork that the doors to the shaft are open and a little dog is playing there with a piece of paper. I am afraid that if the elevator continues downward it may hit the dog, should he happen to have his head or body over the shaft at the time. I try to prevent this.
5/14/48	I pass a number of store windows containing Indian relics.	I am at a show or bazaar. There are booths that display or sell Indian relics and objects.
5/16/48	I am walking along the street with Stella and we come upon a kitten. I touch it gingerly to see if it is alive. It looks at me with sad eyes. I think, "Is the kitten myself?"	(I thought I was awake when this dream occurred.) I lie, having just awakened, on the couch. I feel the soft and gingerly pressure of a cat walking on my body toward my head. I think, "Kurt must have brought one home with him last night!" I lie quietly, waiting for it to reach my neck. As the pressure reaches my neck I get the impression that it is Kurt's head; simultaneously this thing makes a small noise, like the beginning of a purr or a little growl. I gradually realize that I have not been awake. The feeling of pressure on my neck persists slightly even after I awaken. I realize it cannot be Kurt beside me, since I am alone on the couch.

Were it not for the complete breakdown of communication between Kurt and Stella, these dreams probably wouldn't hold much interest—they could be explained too easily by common elements in Kurt and Stella's daytime activity and conversation. As it was, that explanation won't suffice.

Dr. Nelson suggests that the dreams were in fact telepathic, and that they were a way of offsetting the separation Kurt and Stella imposed on themselves during their waking hours. A telepathic contact of this kind, she proposes, "may in itself provide as much and perhaps more pleasure than verbal communication, or it may provide a different kind of gratification which has no representation at the level of conscious awareness." Otherwise, she asks, why should Kurt and Stella, who were at sword's point most of the time, feel gratified—even gleeful—in telling each other about their parallel dreams at breakfast? Why should a man and woman, who were too angry to speak to each other most of the time, get pleasure from the discovery that they were seemingly in close contact during their dreams? Why should they be happy at such a notion, when on the surface they didn't want to be in contact at all? Why, unless the dreams were serving to compensate for their very apartness?

Then there is the case reported by Dr. Margaret A. Paul, a San Pedro, California psychiatrist. In a formal presentation at the Seventh Western Divisional Meeting of the American Psychiatric Association, Dr. Paul told how two of her patients, who were linked to Dr. Paul by strong emotional bonds, both suffered a mysterious loss of memory at the exact time that Dr. Paul experimentally drank a broth made from a hallucinatory mushroom.

Dr. Paul drank the mushroom-based beverage at 8:10 p.m. on a Friday evening, in an effort to find out whether it would produce parapsychological phenomena. While she was under its influence she was whirled through a series of fantasies that included massive destruction by the atom bomb. She woke at 11:30 p.m.

The following Monday, one of her patients, a 30-year-old bachelor, described how he had "lost" three hours Friday evening. He had planned, he said, to watch a TV show at 8 p.m.—but the next thing he knew he was watching the 11 p.m. news. Saturday, he said, he became unexplainably frightened over the prospect of an atomic attack—to the point where he even considered laying in a supply of canned goods.

Had he simply fallen asleep by the TV? It didn't seem so to him. He was sufficiently impressed by the experience to ask several friends if they had seen him in a trance-like state Friday evening, but none had.

Another of Dr. Paul's patients, a 28-year-old unhappily married woman, said she had lost all memory for that same period Friday evening. She had left home about 8 p.m., with another woman, to buy ice cream, she reported, but had no memory of anything that happened until her return about midnight. Imagine her astonishment when her guest informed her that, during this period, she had driven to the home of a married man she secretly loved and sat outside his house mooning over him!

Dr. Paul noted that both patients had unusually strong but confused feelings toward her. Neither patient knew the other. Their loss of memory, she believes, was a response to her own hallucinatory stress.

Does strong emotion, then, facilitate ESP? Can people "pick up" fear, hate, and love telepathically? Experimenters are giving increased attention to this. Their subjects volunteer to undergo routines which, as often as not, put them through an emotional obstacle course designed to swing them from the loftiest heights of patriotism, say, or reverence into the depths of depression and horror.

In Los Angeles, for instance, Dr. Thelma Moss, a psychologist at UCLA's Neurological Institute, has a telepathic sender sit in a darkened room wearing earphones while emotionally loaded pictures are projected on a screen and a cacophony of background sounds are played into his ears. In another room a subject sits on a couch and describes his impressions. One picture shows a violent episode from a Middle Eastern war while shouts, gunfire, and roaring planes besiege the subject's awareness; another shows the Crucifixion, to the strains of Bach's Tocata and Fugue. The subjects report their feelings: fear, horror . . . reverence, bliss. Dr. Moss has obtained results impressive enough for publication in scientific journals not normally given to reports of parapsychological research. In one series of experiments, she found that *artists* get amazingly high scores—especially if they've had previous ESP experiences. Artists who had had spontaneous encounters with telepathy and clairvoyance turned in such a successful performance that the odds against chance were a whopping 6,000 to one!

Strong feelings have a pronounced effect on your body, as you know. Your blood pressure goes up, your breathing and heart rate may increase, you may perspire more, and so on. It stands to reason, then, that if you really can pick up strong feelings telepathically, they ought to be reflected in changes in your body. And they probably are. E. Douglas Dean, a chemist at the Newark College of Engineering, found that changes in blood volume—measured on a device called a plethysmograph—correspond to changes in ESP targets, provided the targets have a strong personal meaning for the subject. The targets Dean uses do have this personal appeal. Dean has his subjects write the names of close friends on cards—names that are important to the subject: perhaps the name of an old girl friend, a hated rival, or a spouse. These, plus an equal number of blank cards, serve as the targets. A telepathic sender looks at the cards, while the receiving subject simply sits with his finger hooked to a plethysmograph. Presumably, when the sender looks at a name that's meaningful to the receiver, the receiver's blood volume will change significantly from what it is when the sender is looking at blank cards.

Dean has tried this over distances as great as 3,000 miles, with the sender in Bordeaux, France, and himself as the receiver in Newark, N.J. The odds against the results being due to chance were 5,000 to one. Dean has gotten such consistently high scores with this method that he hopes it, or a modification of it, eventually can be used to transmit Morse code-like messages. If so, it would be the first practical use of ESP.

What happens to your body, now, if someone you feel really close to—your child or your spouse, perhaps—has an emotionally shocking experience? Does the experience register somewhere in your physical organism, perhaps by ESP? An experiment by five researchers at the University of Montana and Agnes State College suggests that this may be the case.

The research team, headed by George E. Rice at Montana, studied ESP-induced physiological responses between mothers and daughters and between wives and husbands.

In one experiment, the researchers measured changes in the mothers' skin conductivity when a blank cartridge was suddenly fired in front of their daughters some distance away. The skin's electrical conductivity, the well-known "galvanic skin response" (GSR) used on lie detectors, is a subtle measure of changes in perspiration rate—which, of course, is closely tied to the emotions. In this experiment, the mothers' GSR changed appreciably more than that of a similar group whose daughters weren't given the blank cartridge treatment. The mothers were out of hearing of the gunshot, yet their bodies registered the shock being felt by their daughters!

In another experiment, the same researchers measured the galvanic skin responses of wives whose husbands suddenly had their feet dipped in ice water. Again, the GSR changes were greater for wives whose husbands had "cold feet" than for those who didn't—even though the women didn't see what was happening to their menfolk.

These are only a few of the many experiments over the years which suggest, very strongly, that we are never truly isolated from each other. If they are valid, we can only conclude that unknown bonds of thought and feeling reach out from each of us, linking us together in an invisible organismic whole. Clairvoyance, telepathy's twin, seems to suggest that these same mysterious bonds tie us inexorably to the physical universe.

Our friendship and affection for each other is more than a mere subjective feeling. A mother's love for her child is more than just a peculiar chemical state in her body. We cannot say this for a computer or a robot—certainly not for the old "nothing but" man.

The social questions this raises are, of course, enormous. Can hate spread like a disease, flowing from person to person even without the aid of verbal or visual contact? Is that partly why riots, crime, and rebellion have spread so alarmingly in recent years? Are we, literally, in the throes of a "hate epidemic?"

And can love, too, spread? Is that the real meaning of the injunction to return love for hate? Is love the only antidote for our "hate epidemic?" Is love, in a very real sense, a form of social medicine? Perhaps even social nutrition? The evidence that this is the case is now so strong that we can no longer safely ignore it. The implications, not only for our understanding of our own nature but for our social well-being, are overwhelming.

Seeing the Future

A 19-year-old St. Louis girl sat contentedly sipping a soft drink at a high-school dance one night in the mid-1960's. She felt relaxed and dreamily happy.

Suddenly her mood changed. There, on the forehead of the band's Negro drummer, she seemed to see a large letter "D." A moment later it vanished.

"I was so shaken up!" she told me. "I felt sure it could only mean he was going to die."

Two months later that very drummer was brutally murdered in a Chicago gang fight.

Do our minds have the unique ability to travel into the future, actually getting hints and signs of events before they occur?

One day I received a phone call at the *Post-Gazette* from a Cleveland man who told me that his daughter had dreamed the exact number of the Irish sweepstakes ticket she had bought at random the next day. She had mentioned the number at the breakfast table before she bought the ticket, he said. His daughter confirmed this in a separate conversation with me. I noted the number, made a written record of the story, and had the record notorized and placed in the *Post-Gazette*'s office safe. When the sweepstakes winners were announced, I had the record removed. Unfortunately—both for her and for me as a newsman—she did not win. Evidently, her seemingly prophetic dream extended only to that next day. Nor did it have any special importance for her beyond the fact of identifying a ticket. Yet the mere fact of dreaming specifically about a multidigit number before you get the ticket is, to say the least, remarkable.

More personally significant was a series of dreams of Mrs. Cynthia Jacob, a Pittsburgh psychiatrist's wife. The dreams began while she was living in Cleveland, almost exactly one year before she had ever heard of her husband-to-be, Lindsay.

Fortunately, Cynthia was in the habit of writing down many of her dreams. On March 26, 1966, she had the first of what appeared to be several dreams hinting at her forthcoming marriage to Dr. Jacob. She made a written record of it next day. Here are the relevant parts of that dream, taken from her written statement:

"A man came in the front door. . . . He asked me to call him Lindsay or Jacque (Jake). . . . I told him to call me Cynthia. . . . He had penetrating dark eyes, brown-black. I felt very close to him and he to me."

Dr. Jacob's nickname was, in fact, "Jake." Cynthia also got his first name, Lindsay, correctly. Her physical description of him tallied with his real-life appearance.

Cynthia herself says "Jake's" appearance in her dream corresponded perfectly with that of the living Dr. Jacob.

She continued to see him in her dreams. Subsequently

both his first and last names appeared to her in correct order.

On Christmas night, 1966, Dr. Jacob—who, being professionally interested in dreams, occasionally wrote down his own—had a dream which he interpreted as symbolic of his forthcoming marriage (although, being a "happy bachelor" at the time, he had no conscious intention of marrying). The girl in the dream, he says, bore a striking resemblance to Cynthia (whom he had not met). There's one reference to her appearance, in his notes, which bears this out—a comparison with his niece who, in fact, does look like Cynthia.

Cynthia moved to Pittsburgh in the spring of 1967. She met Dr. Jacob in July, 1967, and they were married the following October 9.

Had Cynthia actually met Dr. Jacob at some earlier date and forgotten about it? There's nothing in her own history that would suggest such a thing. Further, Dr. Jacob himself has no recollection of ever having met her. Still, this does not rule out the possibility that they had met and then both completely forgotten each other consciously, retaining buried memories beneath the threshold of awareness—though this possibility is remote.

But no similar explanation can account for the nightmare warnings of Dr. Portia Hamilton, a former New York City psychologist.

Seated in her kitchen in her modest apartment, Dr. Hamilton told me how she desperately tried to save the life of an 11-year-old neighbor, a boy. Her words were chosen carefully. Obviously she was not a crank. Her status as a Fellow of the American Psychological Association testified to that.

For three nights running, Dr. Hamilton had nightmares involving young Dexter Kitson, who lived next door. Each dream showed the child in danger. On the second night, her dreams were so terrible she wanted to get up and warn his family, even though it was after midnight. Her husband had trouble talking her out of it.

On the third night her dreams became unbearable. She resolved that, the next day, she would alert the boy's parents—even though she knew them only through occasional glances out her window.

So on her way to work the following morning, Dr. Hamilton called on the boy's mother, desperately hoping that, somehow, some impending tragedy might be averted.

But the mother did not know Dr. Hamilton. She closed the door in the psychologist's face.

When Dr. Hamilton arrived back home that afternoon she learned to her dismay that the boy had been killed—buried alive digging a cave on a hill some distance away.

Was this precognition—"seeing" the future? Or was Dr. Hamilton clairvoyantly perceiving the boy engaged in dangerous activities?

Precognition would seem the logical explanation. Yet, for many people, precognition is far harder to accept than either clairvoyance or telepathy. Think a moment about what it implies. If people literally do sometimes see the future, they must be perceiving events that are not yet real. It poses a mammoth paradox: how can the human mind, whatever it is, be influenced by something that does not exist? Yet this seems to be exactly what precognition is.

Parapsychologists have their own reasons for being distressed over precognition. Many of them would like to believe in some sort of free choice—as would nearly all of us. But if we can see the future, the future must be spelled out in dreary detail long before we "live" it.

Precognition confounds those who try to find purely physical explanations for ESP. Some people think of telepathy, for example, as a sort of thought wave that travels, TV fashion, from a sending brain to a receiving one. But it would be a strange wave indeed that could travel into the future and return!

When someone tries to tell you ESP is some sort of biological force, ask him what kind of force prophecy is; he'll probably scratch his head and tell you he doesn't believe in prophecy. If he's sophisticated about such things he may tell you precognition is really a sort of elaborate clairvoyance, in which the percipient becomes aware of all the circumstances that are at work on a given person or object at the present time, and then, by extrapolation, can "see" what will happen to that person or object in the future. But this raises philosophical problems of its own, not to mention the tremendous capacities it assumes for clairvoyance.

Take, for example, the results of the standard card-guessing tests for precognition. In these experiments, the researcher tells you to list the order you think the cards will be in after they've been shuffled. Mind you, the cards haven't been shuffled while you're making this list. After you've finished trying to predict the order, the cards are scientifically shuffled. Then the experimenter compares your prediction with the actual order the cards are in.

Such experiments have given significant results time after time over the last 30 years. Assuming the results are valid, how are we to interpret them? Are we to believe that your mind became clairvoyantly aware of the myriad subtle influences that were to affect the process by which the cards were shuffled? Some clairvoyance!

Or take the cases of apparent precognitive behavior in animals. Robert L. Morris, of the Institute for Parapsychology in Durham, made an exploratory study of possible ESP in laboratory rats and came up with what looks much like precognitive behavior—although he himself is cautious in his interpretation of the results. Morris obtained 19 laboratory rats and made careful measurements of their general rate of activity before some were to be killed. A second experimenter, who didn't know which rats had been rated "active" and which "inactive," then killed a selected number of the animals. When the experimenters matched the activity ratings with the list of animals that had been killed, they found that the ones that were killed had been markedly less active during the period before they died than the others—almost as if, anticipating their fate, they had become depressed and lethargic.

If you've ever had to have a cherished pet "put to sleep," you may have seen the slowed-down, depressed-looking behavior such animals sometimes exhibit—almost as if they knew what was coming. That, of course, could be caused by some telepathic influence from your mind to your pet's—or even from some perfectly normal tipoff in your own behavior toward the animal, *or* the animal's own condition. But in Morris' experiment, the condemned animals were chosen at random after the activity measurements had been made. Are we to believe that the animals were able to perceive, clairvoyantly, all the subtle influences that went into the selection process?

My own view of that matter is that the shadow world of time and space melts away when we reach into the deeper levels of our own minds. The distinction between past and future, at that level, is as meaningless and as arbitrary as the distinction between "up" and "down" when you're in outer space. Time and space are fixed entities only in the sensory world of our daily lives. In the strange, little-known realm where ESP seems to function, time and space do not exist.

I recall a particularly striking experience one night in

December, 1952, which brought this sharply home to me. I was 22 years old at the time and was home from college for the Christmas holidays. I had gone to sleep about 10 p.m. An hour or two later I became fully aware of myself and my surroundings but, paradoxically, was unable to wake up. I tried to move but my body failed to respond. I tried to shout but could not. My body seemed as far beyond my control as, say, a table or chair not within reach. It no longer seemed part of "me."

When my initial terror subsided, I had the sudden notion that, in this peculiar state, I might have some unusual psychic abilities. As I recall it, I didn't carefully reason this out, as I've made it sound here. It just sort of "happened." For my target, I chose the weather—the first thing that occurred to me, I suppose. I became aware, in a dreamlike fashion, of a cold wave about to move toward my home in Buffalo, N.Y., from Colorado. Moments later I regained my normal consciousness.

That morning, the weather forecaster announced that, as I had apparently foreseen, a cold wave was headed for the northeast from Colorado. I hadn't seen any weather maps the night before, nor had I any reason to believe Colorado was spawning a cold wave for us. If I had made a guess as to where a possible cold air mass might be coming from, I'd have said Montana, the Dakotas, or Minnesota—certainly not Colorado. Yet that's where the weather man placed it—and that's where I had located it in my dreamlike condition.

Was this clairvoyance or precognition? As I look back on it, there seems to be no way of telling. And that's the strange part of it. Normally you can tell whether you're seeing an event at another point in space or recalling it from a different point in time. But I seem to have been unable to make this distinction. Does this mean that, for the ESP-attuned mind, there is neither space nor time, and that such arbitrary classifications as "precognition" and "clairvoyance" are meaningless? I believe it does.

This was also suggested, I think, in one of the early pioneering experiments in ESP by the noted British psychical researcher, Whately Carington. Shortly before the Second World War, Carington carried out a series of experiments in Cambridge, with pictures as the targets. He was looking, however, for evidence of telepathy, not precognition. On each of 10 evenings, he hung up in his study a picture of a familiar object. He asked a number of people scattered throughout England, America, Holland, and Scot-

Seeing the Future

land to try to reproduce, by drawing, from telepathy, whichever picture was hanging in his study at the time. (The "recipients" were not familiar with the pictures.) The drawings and target pictures were then matched and scored by a third person who did not know which drawing had been made on which night.

There was no significant resemblance between the drawings and whichever picture was hanging up at the time. But there were striking similarities between the drawings and the pictures that were going to be used on the next two evenings! There was an equal tendency for subjects to draw pictures like the ones that had been used the previous two evenings. The odds against the results being due to chance were calculated at 10,000 to one. Apparently the percipients' minds made no distinction between past, present, and future. They were trying for a drawing that was hanging on the wall at the "present" time! This suggests, again, that the mind makes no distinction between telepathy, clairvoyance, and precognition. Space and time are all one—or, rather, all nothing—to the deeper unconscious mind.

In 1939 a mathematics lecturer at London's Queen Mary College, Dr. S. G. Soal, finished an elaborate attempt to repeat the card-guessing experiments pioneered by Dr. Rhine. Tests on 160 people had yielded only chance results. Soal publicly expressed skepticism over Dr. Rhine's work. But Carington persuaded Soal to re-examine his records with an eye to finding out whether his subjects, like Carington's, hadn't been displacing their calls in time. Sure enough, Soal found that two of his subjects had been calling—correctly—the cards just before and just after the target card. This, of course, wasn't necessarily precognition—the cards had already been arranged in order, and all the subjects' minds had to do was jump a little further along the existing deck.

But one of these subjects, Basil Shackleton, showed an amazing ability to continue this strange displacement even after random number tables were brought on the scene to determine the order of the cards. As Shackleton tried to guess each card the sender was looking at, the next card had not yet been chosen—and wasn't chosen until after his guess had been made, at which time the sender was shown a number from the random numbers list, which in turn was used to choose the next card.

Shackleton's unusual performance continued. He persisted in naming the card that was *about* to be selected.

But clairvoyance still hadn't been completely ruled out —after all, didn't the random numbers table exist in the present? Couldn't his mind have somehow perceived *it* directly, instead of actually looking into the future?

So Soal and his assistant, Mrs. K. M. Goldney, replaced the numbers list with a bag of 200 colored chips. One chip was plucked from the bag by Mrs. Goldney just before each card was chosen. The color of the chip was used to decide which card would be selected. Shackleton continued his off-beat performance. Mrs. Goldney even went so far as to try to avoid any fixed habits in choosing chips from the bag so Shackleton couldn't clairvoyantly learn these habits and, by inference, tell which card was coming up next.

Then an even more striking feature emerged. On most of these tests Shackleton guessed the card the sender was going to look at two or three seconds later. At the normal guessing rate of one call every two to three seconds, this meant he was guessing one card ahead. But, amazingly, when the speed of the guessing was doubled, Shackleton's scoring shifted to the card two places ahead!

With three different senders, Shackleton totaled about 7,000 guesses at the normal speed. His scoring on the card one place ahead averaged seven per run—in contrast to the 4.8 that would be expected by chance (4.8 instead of 5 because there are only 24 opportunities per run of 25, of guessing a card one place ahead). The odds against explaining these results in terms of chance are astronomical.

Again we see evidence of the mind operating in a kind of spaceless, timeless never-never land where "here," "there," "now," "tomorrow," and "yesterday" do not exist.

Over and over, experimenters have gotten results in precognition tests which excluded chance by overwhelming odds. Dr. Margaret Anderson, a psychologist currently with the Mt. Lebanon High School in Mt. Lebanon, Pennsylvania, originally undertook experiments at Duke University's Parapsychology Laboratory because, as a teacher, she felt ESP might play a part in communication between teachers and their students. Aided by Mrs. Elsie Gregory, a Wheaton, Illinois, classroom teacher, she carried out a series of precognition tests with a sixth-grade class which, over a period of a year, gave results that chance couldn't explain one time in 10,000.

In still other precognition experiments, Michael Sanders at Duke obtained results that ruled out chance by odds of 1,250 to one, and Dr. John Freeman, also at Duke, re-

Seeing the Future

ported scores that couldn't be explained by chance one time in 1,000 (although his subjects turned in only a chance performance on a *clairvoyance* test).

Parapsychologists have tried to rule out clairvoyance with all sorts of wild, imaginative gimmicks. They've used rolling dice to decide the order of cards. They've used machines to shuffle the cards. They've used machines to roll the dice that decided the order of the cards. They've developed coding procedures based on information to be published in future editions of newspapers. But the high test scores have continued to come in.

For people who believe in precognition in the first place, of course, this is all just so much monotonous claptrap. Haven't we had enough proof from "prophets" like Washington seeress Jean Dixon, medical clairvoyant Edgar Cayce, and crime-solver Peter Hurkos—all of whom have, at various times, put forth predictions that couldn't be explained by chance? Predictions, moreover, that had a vital impact on the lives of the people involved?

True, these gifted people have done remarkable things. But honest scientists must remain skeptical of this kind of material for one excellent reason: people have an annoying but almost irresistible tendency to remember the successful predictions and forget the misses. Consequently, the reported hits pile up in impressive majesty, while the failures get swept into the trash baskets. The result is a persuasive but misleading impression of near-infallibility.

Take Edgar Cayce. The famed medical seer, whose clairvoyant career was studded with many phenomenally successful readings, undoubtedly did great good. In a trance-like state, he was apparently able to give remarkably precise diagnoses of a wide range of mostly distant and consciously not-known-to-him disease conditions in people he'd usually never met, although he had no medical training himself. His over-14,000 readings have been faithfully preserved at Virginia Beach, Virginia, where the organization dedicated to his memory—the Association for Research and Enlightenment—maintains its headquarters.

The thousands of people who felt that they were helped by this kindly, honest man became so impressed by his medical insights that, over the years, they quite naturally sought his advice on other matters.

Cayce was a firm believer in free will. Often, when he was asked about future events, he replied that a great deal depended on the attitudes and behavior of the people involved. Would there be a war? It depended on whether men were selfish. What would be the future of Hitler's Germany? Much, again, depended on whether men allowed selfishness to take precedence over the purposes of the Prince of Peace. All very true, perhaps, but hardly a prediction of things to come. He drew a clear miss on China, which he saw becoming democratized. He also predicted a depression for 1956—another miss. He saw friendship between Russia and the United States. ("By what will it [Russia] be guided? By friendship with that nation which hath even placed on its monetary unit, In God We Trust.")

On the other hand, there were those who grew wealthy by following his predictions. In the 1920's he advised buying land in areas of Virginia Beach which were destined to skyrocket in value—although no one knew it at the time. He also foresaw a tidewater boom which netted one Virginia Beach businessman an eight-fold profit on a $10,000 80-acre tract.

Probably his best-known prophecies deal with extensive and tragic geological changes. He predicted widespread upheavals from the 1960's into and beyond the year 2,000. If Cayce is correct, much of the East Coast will be drowned under the rolling Atlantic. The western United States, too, will become a watery disaster area—sooner than the East. The only regions that will escape the onslaught are parts of Ohio, Indiana, Illinois, and southern and eastern Canada.

There is no doubt about one thing—Cayce predicted upheavals and earthquakes in many parts of the world where, geologists say, unsettled earth conditions do in fact exist. His predicted areas of disturbance do seem to match well-defined earthquake zones. But is this truly precognition, or had Cayce absorbed some geological information from casual reading? His followers rule out the latter—not only was he not prone to read about such things, they say; geologists didn't even discover some of the conditions he envisioned until years after the readings had been given.

The little-educated Cayce, who had performed seeming wonders as a medical diagnostician, was asked to perform equal wonders in settling many of life's major philosophical problems. His predictions were—and are—considered almost infallible by many people. My tone here doesn't mean, of course, that all his predictions were wrong or that he had no unusual precognitive abilities. It merely tries to show how people often tend to over-generalize about their heroes—when they see someone turn in a stellar performance in one field, they expect him to turn in stellar performances in everything. They compile their hero's successes and forget his failures. It's an easy mistake to make, and it's why parapsychologists are cautious about men like Cayce and Peter Hurkos.

Hurkos, the noted clairvoyant who was brought in on the Boston Strangler case, turned in one set of predictions back in 1959 which provide clear evidence of his prophetic abilities—and limitations. At the start of the baseball season, he visited both the major league training camps in Florida. The Dutch psychic knew virtually nothing about baseball. He did not talk extensively to managers or players. Typically, he merely strolled onto the playing field, perhaps touching the manager gently on the shoulder, perhaps fondling a bat briefly and noncommittally.

His understanding of baseball—or rather lack of it—was revealed in his occasional accidental references to it as "football" and in his comment to the Chicago White Sox manager, Al Lopez:

"When you see somebody with a stick knock with the ball, he runs away like crazy and then another guy runs to another corner. For what?" Lopez reportedly moved quietly away.

At the end of his tour, Hurkos gave his predictions for the standings in both major league clubs. His prognostications were published in *This Week* magazine April 26, 1959.

For the National League, he saw it this way:

1. Los Angeles Dodgers
2. Milwaukee Braves
3. Cincinnati Redlegs
4. Pittsburgh Pirates
5. Chicago Cubs
6. San Francisco Giants
7. St. Louis Cardinals
8. Philadelphia Phillies

The actual standings at the end of the season:

1. Los Angeles
2. Milwaukee
3. San Francisco
4. Pittsburgh
5. Chicago

6. Cincinnati
7. St. Louis
8. Philadelphia

Except for one transposition—switching Cincinnati and San Francisco—Hurkos hit the occult equivalent of a grand slam home run. Skeptics will probably argue that he had secretly acquired a knowledge of baseball, and that his predictions were just well-informed guesses. If he did acquire such knowledge, this was a strange way to display it, since most sports experts and informed fans had picked Los Angeles to end up around seventh place.

For the National League, Hurkos had done better than the computers which *This Week* magazine had traditionally used to predict the pennant race. In the words of writer Leslie Lieber:

"We've fed them [computers] a bleacherful of baseball statistics—but in each case the machine struck out, dismally."

Hurkos was much less successful for the American League. His predictions:

1. New York Yankees
2. Chicago White Sox
3. Cleveland Indians
4. Boston Red Sox
5. Washington Senators
6. Detroit Tigers
7. Kansas City Athletics
8. Baltimore Orioles

The actual standings:

1. Chicago
2. Cleveland
3. New York
4. Detroit
5. Boston
6. Baltimore
7. Kansas City
8. Washington

In other words, he blew it. Oh, not completely. He had made special mention of Kansas City as ending up in second to last place. Baltimore, to Hurkos, was the Grapefruit League lemon—but the team didn't end up quite as far down as he predicted.

Hurkos assured Al Lopez the Chicago club couldn't end up lower than second place—which was correct, although the team won the American League pennant instead of settling for the second place Hurkos had given it in his ranking.

He was also right in placing Cleveland and New York among the top three.

But let's face it from a practical standpoint, he blew it. If your bets had been placed on Hurkos' American League predictions, you would have emerged less than rich.

Yet if you had bet the Hurkos way on the National . . .

His National League predictions were little short of uncanny. They simply can not be explained as coincidence. Had it not been for the Redlegs-Giants reversal, he'd have gotten a perfect score—and even that unlucky switch might be explained by the fact that Cincinnati was one of the two teams Hurkos didn't get the "feel" of personally: he had to be content to sit in their locker room, sensing the surroundings, while the Redlegs were playing miles away (the other team he couldn't contact personally was St. Louis).

What else could this have been but precognition? He got far more than one "hit" in the National League instance. It certainly doesn't seem likely that the National League players all read the initial predictions in *This Week* magazine and then deliberately played in a way that would land them in the predicted order.

Fraud on the part of Hurkos or *This Week* is out of the question, of course, since his predictions were published months before the season was completed. They were read by thousands of people on April 26, 1959. They are a matter of public record. You can look up that issue of *This Week* yourself and see the predictions. Then you can look up the actual standings in a 1960 World Almanac. How could a thing like that have been faked?

The moral, of course, is that Hurkos, like Cayce and other gifted persons, can give truly remarkable predictions —and can make spectacular blunders. While the achievements of these people may add to the evidence for precognition, they are not dependable enough to have practical value.

But even if precognition has little practical value so far, it has tremendous value for our understanding of ourselves. Its implications go beyond those of telepathy. If precognition is a real phenomenon, it proves—firmly and finally—not that man has some sort of "extra sense," but that man's very relationships to time and space are different from those of, say, a computer or a clever robot.

We are not, if precognition is a fact, the "nothing but" men of the iconoclasts. We are not limited to the here and now. We transcend our physical selves. The notion of prophecy revolutionizes our ideas about ourselves. If precognition exists, we can only conclude that some deep, essential part of ourselves transcends the day-to-day world of common experience and exists, in a very real sense, in eternity.

The accumulated evidence for precognition leaves little doubt as to its reality. The odds that chance could explain the results of the many experiments that have been performed are *one in thousands*. The nature and extent of precognition are still to be defined, but the fact that something mysterious happens in these experiments cannot be denied. The evidence for precognition and other forms of ESP has been made even more impressive by hypnotists' efforts to reproduce the phenomena under trance.

* VI *

Hypnosis and ESP

A 75-year-old Pittsburgh woman lay dangerously ill on her living room sofa one Saturday afternoon in 1961. Her son, Arthur M. Lawrence, sat beside her, holding her hand and trying to comfort her.

A few days earlier the arteries in Mrs. Lawrence's aging temples had become severely inflamed. They had swelled alarmingly, the blood flow was markedly reduced, and the pain was excruciating. Hot packs had to be kept on her head. Her fever sometimes reached 105°—a life-threatening level in an elderly woman.

As her son sat quietly beside her, she suddenly began moaning in distress. "Are you worse, mother?" Mr. Lawrence asked.

Apparently speaking in delirium, Mrs. Lawrence told her son that she was standing at the top of a high cliff, a strong wind was blowing, and she was terribly frightened. At any moment, she thought, the gusts would carry her off.

Mr. Lawrence responded almost intuitively. "I sensed that the cliff symbolized death," he recalled later. "I was afraid that if she felt she was actually blown over, she would literally die."

In a firm voice, Mr. Lawrence told his mother not to be afraid. Then he added:

"I'm building a high strong fence around you, and nothing can possibly harm you no matter how fiercely the wind blows."

He reinforced this with more words of assurance. Then very quickly he told her the fence was finished.

Her moaning decreased, he recalled, and she again seemed peaceful.

Several months later, after Mrs. Lawrence was out of danger, she told her son she had almost died.

"I felt I was being blown over the side of a cliff," she declared.

"What saved you?" he asked.

"You built a fence," she said.

This is a first-rate example of hypnosis. In answer to her son's suggestion, Mrs. Lawrence had "built" a fence in her mind—a fence which, however, was so real to her that it brought her relief during an extremely frightening situation.

Who knows? It may even have saved her life. Your thoughts and your health are so closely linked that, when you're on the edge of death, your own state of mind may determine which way you go.

The amazing thing about this story is that Mr. Lawrence is not a hypnotist! He had no thought of hypnotizing his mother. He simply gave her a strong suggestion at a time when she was most receptive. Yet this is exactly what hypnosis is. The usual way of hypnotizing someone involves just this pattern. You approach a mind that's in an open, uncritical, un-analytical state and impose strong suggestions on it. You repeat, over and over, words and phrases that conjure all the warmth, looseness, limpness, drowsiness, and sleepiness language can project. Soon your subject's eyes close. You give him suggestions aimed toward controlling major body muscles. You direct him to straighten his arm, then tell him he can't bend it. Assuming he's in a light trance—and, if your suggestions have worked so far, he will be—that arm will stay stiff, no matter how he seems to struggle. Then, to restore his arm to normal, you simply tell him he can bend it again.

From there you proceed to larger muscle groups, then to simple illusions—such as the feeling of floating—and finally to more complex illusions, including complete hallucinations. At that point, your subject will be in about the same state of receptivity Mrs. Lawrence was in. Basically, that's all there is to hypnosis.

Not everyone, of course, can be put into a deep hypnotic trance. Estimates vary, but most authorities will tell you that only about one person in five can be hypnotized deeply enough to "see things that aren't there" on command.

Simple as it is, hypnosis is as strange, as little-understood by science, as it was when it was first practiced by Anton Mesmer nearly 200 years ago. Not even our most learned scholars truly understand the change that takes place. Terms like "hyper-suggestible" do little to clarify it. Why will a good hypnotic subject forget everything that happened during a hypnotic session if you, the hypnotist, tell him to? And then remember it, if you tell him, during a second session, to do so? By what mechanism does a deep-trance subject conjure up animals, people, far-away scenes—anything the hypnotist suggests? And not see anything the hypnotist tells him isn't there? By what means, paranormal or otherwise, do post-hypnotic orders become implanted in a subject's mind so that, at a specified time days or weeks later, he carries them out precisely?

Is it merely some mechanical process, like programming a computer? But how, in computer technology, could one describe the difference between being hypnotized and not hypnotized? Does it make sense to talk about a computer being in a highly suggestible state?

Even if hypnosis itself could be explained in terms of known scientific principles, a hard core of phenomena elicited under hypnosis could not.

Almost from hypnosis' beginnings, bizarre effects have, on occasion, been attributed to it. The Russians have relied heavily on it in much of their research on telepathy —as we will see in a subsequent chapter. From the earliest records of the British and American Societies for Psychi-

Hypnosis and ESP

cal Research, we have stories like the following, reported by a 19th-century physician and verified through interviews with the participants. A doctor in England had experimented extensively with a woman who claimed the ability to "travel" clairvoyantly under hypnosis, perceiving people and events in distant places. The doctor writes:

> I told a patient of my own, Mr. Eglinton, at present residing at the village of Tynemouth, that I intended to visit him [via the clairvoyant]. He stated that he would be present between 8 and 10 p.m. in a particular room, so that there might be no difficulty in finding him. He was just recovering from a very severe illness, and was so weak that he could scarcely walk. He was exceedingly thin from the effects of his complaint.
>
> After the usual state had been obtained, I said, "We are standing beside a railway station, now we pass along a road, and in front of us see a house with a laburnum tree in front of it." She directly replied, "Is it the red house with a brass knocker?" I said, "No, it has an iron knocker." I have since looked, however, and find that the door has an old-fashioned brass handle in the shape of a knocker. . . . [They moved along through the house to the room where he thought Mr. Eglinton would be.]
>
> After a little she described the door opening, and asked with a tone of great surprise, "Is that a gentleman?" I replied, "Yes; is he thin or fat?" "Very fat," she answered; "but has he a cork leg?" I told her that he had no cork leg, and tried to puzzle her again about him. She, however, assured me that he was very fat and had a great corporation. She also described him as sitting beside the table with papers beside him, and a glass of brandy and water. "Is it not wine?" I asked. "No," she said, "it's brandy." "Is it not whisky or rum?" "No, it is brandy," was the answer; "and now," she continued, "the lady is going to get her supper, but the fat gentleman does not take any." I requested her to tell me the color of his hair, but she only answered that the lady's hair was dark. I then inquired if he had any brains in his head, but she seemed altogether puzzled about him, and said she could not see any. I then asked her if she could see his name upon any of the letters lying about. She replied, "Yes;" and upon my saying that the name

began with E, she spelt each letter of the name "Eglinton."

I was so convinced that I had at last detected her in a complete mistake that I arose, and declined proceeding further in the matter, stating that, although her description of the house and the name of the person were correct, in everything connected with the gentleman she had guessed the opposite from the truth.

On the following morning Mr. E. asked me the result of the experiment, and after having related it to him, he gave me the following account: He had found himself unable to sit up to so late an hour, but wishful fairly to test the powers of the clairvoyant, he had ordered his clothes to be stuffed into the form of a figure, and to make the contrast more striking to his natural appearance, an extra pillow pushed into the clothes so as to form a "corporation." This figure had been placed near the table, in a sitting position, and a glass of brandy and water and the newspapers placed beside it. The name, he further added, was spelt correctly, though up to that time I had been in the habit of writing "Eglington," instead of as spelled by the clairvoyant, "Eglinton."

Granted, this is far from typical. Attempts to induce hypnotized people to report distant events usually fail. The subject may give a detailed, convincing description of whatever you tell him to "see," but if you check it you'll usually find his account was wrong. There's a simple enough reason for this: a deeply hypnotized person will try to satisfy the hypnotist, even if, to do so, his unconscious mind has to fabricate scenes, visions, voices, and events. Hypnotized subjects have been told to "visit" the moon and have described an environment much different from the one our astronauts actually found.

That, at least, is what happens in run-of-the-mill cases. But it's not the run-of-the-mill cases that interest parapsychologists. The Eglinton case and others like it have never been satisfactorily explained. Nor has anyone explained how, on rare occasions, some hypnotized subjects enter into such close rapport with the hypnotist that they have apparently been able to share the very feelings and sensations being experienced by the hypnotist!

One of the earliest such cases was recorded in the Proceedings of the Society for Psychical Research late in the

Hypnosis and ESP

last century. Edmund Gurney, a founder of the society, made a careful study of a mysterious bond between a hypnotist, G. A. Smith, and his subject, identified as "Conway," on Sept. 10, 1883, in England. Gurney's report of the experiment:

> Conway sat with his eyes closed in a tolerably deep trance. Mr. Smith and I stood beside him, without contact, and Mr. Smith preserved absolute silence. I from time to time asked Conway whether he felt anything, but of course gave no guiding hint or indication of whether he was right or wrong.
>
> I pinched Mr. Smith's right upper arm. Conway at once showed signs of pain, rubbed his right hand, then passed his left hand up to his right shoulder, and finally localized the exact spot.
>
> I silently changed to Mr. Smith's *left* arm. In a very few seconds Conway's right hand flew to the corresponding place on his own left arm, and he rubbed it, uttering strong complaints.
>
> I nipped the lobe of Mr. Smith's right ear. Conway first rubbed the right side of his neck close to his ear; he then complained of his right leg, and used threats. I then gave a severe nip to his *own* right ear, and he made no sign of any sort. He then rubbed close to the left ear, and finally localized the spot on that ear exactly corresponding to the place touched on Mr. Smith's right ear.
>
> I now pinched the right side of Mr. Smith's right thigh. Conway, without receiving any hint that he was expected to feel anything, immediately began to rub the corresponding part of his *left* leg.
>
> Mr. Smith now put a succession of substances into his mouth, according to my indications, still keeping behind Conway, and preserving total silence. I kept Conway's attention alive by asking him from time to time what the taste was like, but gave not the faintest guidance, except in the case of cloves, when—to see if Conway would take a hint—I asked if it tasted like spice, and he said it did not.
>
> *Mustard*—"Something bitter . . . It's rather warm."
> *Cloves*—"Some sort of fruit . . . Mixed with spirits of wine . . . Not like spice . . . Tastes warm."
> *Bitter Aloes*—"Not nice . . . Bitter and hot . . . Sort of harshness . . . Not sweet." (I had suggested

that it was sweet.) "Not nice . . . Frightful stuff . . . Hurt your throat when you swallow it . . . Bitterness and saltiness about it."

Sugar—"It's getting better . . . Sweetish taste . . . Sweet . . . Something approaching sugar."

Powdered Alum—"Fills your mouth with water . . . Precious hot . . . Some stuff from a chemist's shop, that they put in medicine . . . Leaves a brackish taste . . . Makes your mouth water . . . Something after the style of alum."

Cayenne Pepper—Conway showed strong signs of distress. "Oh! You call it good, do you? Oh! Give us something to rinse that down . . . Draws your mouth all manner of shapes . . . Bitter and acid, frightful . . . You've got some cayenne down my throat, I know." Renewed signs of pain and entreaties for water.

The subject was now awakened. He immediately said, "What's this I've got in my mouth? Something precious hot. Something much hotter than ginger. Pepper and ginger."

A hoax? Hypnotists are often accused of fakery, perhaps because of their association with stage magic. They are more vulnerable to this charge than other researchers. But if the foregoing was a hoax, how does one explain the remarkable results obtained by Lawrence Casler in careful, laboratory experiments at The City College of the City University of New York in 1961 and 1963?

A typical experiment involved 15 deeply hypnotized college students—nine men and six women—who were instructed to try to tell which of 25 cards a hypnotized woman was looking at in a nearby room.

When Casler compared their performance while they were hypnotized with their performance on the same test in the normal waking state, he found that all but four of the subjects consistently got higher scores when they were hypnotized. The combined number of hits for all subjects under hypnosis was 326, in contrast to a combined total of 283 when they were awake—out of a possible total of 1500. The odds against this difference being due to chance alone are 100 to one.

Hypnosis has given us what may be the nearest thing we have to a truly repeatable card-guessing experiment. A Czechoslovakian parapsychologist, Milan Ryzl of Prague, used hypnosis to develop ESP ability in a Czech librarian, Pavel Stepanek.

In guessing the color of cards sealed in opaque envelopes, Stepanek scored far above chance repeatedly, over a period of thousands of trials, through 1962 and 1963. This was one of the few cases where positive results were achieved so consistently that, without question, the experimenters could invite other scientists to witness demonstrations with no fear that Stepanek's scoring levels would fall off.

But all good things apparently must come to an end—and they have a way of doing so in parapsychology with distressing frequency. By 1964 Stepanek's scores had declined to chance level. For four long years, efforts to restore his performance to its old standard met with only sporadic success.

Then, working with Stepanek at the University of Virginia, Dr. J. G. Pratt—who had cooperated with Ryzl—began getting promising results again. But in a rather different way. Stepanek was no longer getting his usual high percentage of right answers. What he was doing, quite unexpectedly, was giving the *same* guesses, right or wrong, for the *same* cards, over and over! The cards, sealed in envelopes, were not visible to Stepanek. This tendency to give the same guess for a given card repeatedly, trial after trial, has been called the "focusing effect." It commands a major share of the attention of researchers working with Stepanek today.

Meanwhile, up in Finland, a Helsinki psychologist—Jarl Fahler, then president of Finland's Society for Psychical Research—began getting spectacular ESP scores in card tests with one of his hypnotized patients. Part of his success may have grown from the close doctor-patient relationship he was able to develop with her. Aided by his skillful hypnotic treatments, she recovered from a severe illness and was able to work and marry.

In experiment after experiment, Fahler's hypnotic suggestions brought out striking ESP scores. Between 1957 and 1963 he published five articles describing his remarkable work in scientific journals. His reputation grew among parapsychologists. In 1965 he and Dr. Karlis Osis of New York, who established a formal research department within the New York-based Parapsychology Foundation, carried out one of our most important probes into the mysteries of hypnosis and precognition.

Their aim was to find out how reliable a person's

hunches are. You know, of course, what it's like to have a strong feeling that a certain event will soon take place—a hunch that a certain person will call today or that a specific letter will arrive in today's mail. Well, surprisingly enough, many of our precognition experiments don't allow for *hunches* at all! In most, the subject simply tries to guess the cards that will turn up after the deck has been shuffled. The experimenters seldom consider how the subject feels about the rightness or wrongness of a given guess.

Fahler and Osis tried to take this into consideration. Fahler hypnotized two subjects, chosen especially because they believed they sometimes could tell when their ESP guesses were right. While they were hypnotized, Fahler gave them suggestions to reinforce these hunches. On certain guesses, he told them, they could expect strong impressions of correctness. The subjects were told to say "mark" each time they made a guess they felt was right.

Instead of cards, the experimenters used the numbers from one to 10 as targets. The subjects were asked to guess which numbers would later be written on the record sheets. (Fahler and Osis didn't choose the numbers until later; they were selected by an elaborate randomizing procedure.)

Between them, the subjects made a total of 1,950 guesses over a period of 13 sessions.

The results were almost breathtaking. When Fahler and Osis ran a statistical analysis on the total number of correct guesses, they found nothing that couldn't be explained by chance; if anything, the scores were slightly below chance levels. But the story was different for the guesses the subjects had *felt* were correct. Taking only those into account, the experimenters found results for which the odds against chance were fifty million to one!

A hoax? Did Fahler and Osis cheat by selecting only subjects who *happened* to get high scores through chance alone, ignoring all the others? This, as I said earlier, is a common criticism many scientists level against parapsychology. But to pull off that little stunt, logically, Fahler and Osis would have had to run millions upon millions of experiments.

Hypnosis is playing a vital role in unlocking the secrets of the weird and wonderful world of dreams. This, of course, is because you can dream when you're hypnotized. What's more, you can dream about anything the hypnotist suggests to you. You can dream, and use symbols in the

Hypnosis and ESP

content of your dreams, much as Freud and Jung conjectured. Thus, for example, one deeply hypnotized youth, who was taking part in a study of the symbols dreamers use to disguise a deeply buried desire to regress back to the womb, was told:

"You will dream you have lost your job and need money, and that you escape from the situation by going back into your mother's womb."

His dream, as he reported it moments later:

"I dreamed that a mouse was running through thick grass and I think I was the mouse. Got to some water and the mouse had to swim across the water and kept on running."

Another subject, given the same suggestion, had the following dream:

"I am walking down a tunnel which is quite brilliantly lighted, although there are no bulbs evident. It doesn't seem that I can ever reach the end."

The grass, the water, and the tunnel would, according to Jungian psychoanalytic theory, represent a return to the womb, with the mouse possibly representing the fetus.

Contrast these dreams with the dream had by this last subject in response to the following suggestion:

"You will dream you have lost your job and need money. You escape from the situation by becoming a little boy again, by going back to the time when you were only 10 years old."

Dream: "A little boy is in a dime store and he is reaching up to steal some play money. He takes it off the shelf and puts it under his coat and walks out of the store."

All these dreams, which were part of a formal experiment in dream symbolism I carried out at the University of Montana, were experienced and reported by the subjects within a very short time. All the subjects were deeply hypnotized and all had dreams which strongly resembled the natural dreams we have at night. The advantage of using hypnosis is that you can speed the whole process, and that you know in advance exactly what stimulus evoked the subject's dream.

At the dream laboratory in Brooklyn's Maimonides Hospital, Dr. Stanley Krippner has been using hypnosis to study dreams. This is really a refinement of the work by Ullman and Krippner which we described in Chapter IV. Instead of letting all the subjects fall asleep naturally in the dream lab, however, Krippner hypnotized one group and compared their dreams with the dreams of another group—a group that was relaxed but not formally hypnotized.

The hypnotized subjects were told that telepathy often occurs during hypnosis and were instructed to try to imagine which of several pictures was being viewed by a sender in another room.

The subjects who weren't hypnotized were given similar instructions, except that all references to hypnosis were omitted—references to "imagination" were used instead.

All the subjects were told they would have a chance to take a 45-minute nap after the session, and that they might then pick up more images—telepathically, of course —of target pictures being viewed by the sender.

Finally, they were told that they might dream about the target pictures during the coming week, and to report their dreams to the experimenters seven days later.

Their dreams were written down; three independent judges tried to match the dreams with the appropriate pictures.

The judges were able to match the hypnotized subjects' dreams with the pictures to a highly significant degree for the period in which the subjects were napping—the odds against the results being due to chance were a thousand to one.

The judges were also able to match up the non-hypnotized subjects' dreams and the pictures—but only for the week when the subjects had left the laboratory and were dreaming at home. In this case, the odds against chance were 100 to one.

Dr. Krippner's interpretation may throw light on the role of hypnosis in telepathy generally. Perhaps, Dr. Krippner suggests, hypnosis speeds up the processing of telepathic material. Perhaps, he says, the non-hypnotized subjects required more time for the telepathic material to reach consciousness. But much more research is needed to confirm this.

Hypnosis has not been popular among American parapsychologists. It has been used far more extensively by the Russians in their studies of telepathy. But that's a subject for another chapter (Chapter X).

* VII *

Mind over Matter

A strange psychic force launched an outlandish display of mischief one winter night in 1933 in an old house in Rovno, a city of about 45,000 in eastern Poland. Sergei and Olga Aposhansky and their children were seated in the dining room entertaining an overnight guest. Their 23-year-old housekeeper, Nina Rudnicki, lay sleeping in her bed in the kitchen, as was the custom in Poland at that time. At about half past midnight a dishpan was heard sliding from its appointed location at the center of the kitchen table and clattering to the floor.

Two hurried strides carried Mrs. Aposhansky into the kitchen. There on the floor lay the rebellious dishpan. And there in her bed, some distance away, lay the soundly sleeping Nina. Mrs. Aposhansky replaced the dishpan and rejoined her company.

Almost at once the pan fell again.

The guest, an impetuous fellow, volunteered to put it back, announcing as he did so that this time it would certainly stay put! He placed it squarely in the center of the table. No sooner had he returned to the dining room than clatter, crash, the perverse dishpan repeated its annoying trick! Nina slumbered on.

There followed an onslaught of flying boots and shoes. As the events were reconstructed for me by Catherine Aposhansky, who was 19 years old when the strange force invaded her parents' home, the objects seem to have just "taken off" from their positions on the floor and flown about 12 feet, landing with great force.

The story, which she had witnessed herself, wound through the byways of a frustrated love affair and culminated in an eerie demonstration of apparent ESP under hypnosis.

Boots, shoes, Nina's purse, and the reprehensible dishpan continued their unwelcome migrations nightly for about three months. Nina, of course, was an object of the gravest suspicion by members of the family. Didn't the strange force always manifest itself in her presence? Weren't most of the flying objects located in the kitchen? Wasn't she the one—the only one—who had actually seen a box of matches fly off the kitchen table, circle over Mr. Aposhansky's head three times while he was cleaning his teeth, and land on the floor behind him? Suspicious stuff indeed!

They decided to test the suspect Nina. High above the kitchen stove was a shelf. It was much too high for anyone to see over just by craning his neck. You had to stand on a chair to put anything on it. This shelf, then, would serve as an ideal proving ground for the mysterious Force.

So one day when Nina was away, one of Catherine's brothers placed a small bell on this shelf. That night the unsuspecting Nina went to bed as usual. And, as usual, objects began flying noisily around the darkened kitchen. About 12:30 a.m., after the boots, shoes, purse, and dishpan had had their say, the bell mysteriously started ringing. Then it, too, jangled to the floor. Nina had been vindicated.

Then things began happening during the day. The same objects began flying at times when Nina was awake and under the full observation of those present. Other articles soon joined the fun. There was the time, for instance, when a milk delivery woman was standing unsuspectingly in the kitchen waiting to be paid, and, before her startled eyes, a cup suddenly took off from a table, streaked through the air, and crashed at the poor woman's feet. The hapless milk woman vowed she would never again return to that house.

And there was the time an electric light, suspended from the kitchen ceiling by a four-foot cord, began—in full view of the assembled family members—to swing in circles that slowly grew larger and larger until the light was arcing around in a five-foot orbit. Then, smoothly and evenly as if by some intelligent plan, the circles grew smaller and the light returned to its stationary pose.

At odd times during the day and night, loud snapping noises, which Catherine Aposhansky compared with a man snapping his fingers "only much louder," issued disconcertingly from somewhere near the kitchen ceiling.

Mind over Matter

By now the entire Aposhansky family had stopped thinking of the mysterious force merely as "it." The force had become "him." "He" had taken on a personality and individual identity of "his" own.

"Let's see if he can play the guitar," a young Aposhansky suggested.

So they left a stringed instrument—roughly the equivalent of one of our guitars—in one of "his" favorite haunts in the kitchen.

Sure enough, "plunk, plunk, plunk." "His" feeble attempts seemed to acknowledge that, even if he lacked musical ability, he at least had a sense of humor. Hesitantly —they were all a bit afraid of "him"—they went into the kitchen. On retrieving the instrument, they found long gouged-out grooves in the wood, as if some superhuman pair of fingernails had had unwelcome access to it.

Word of the Aposhansky's strange visitor spread far and wide through the district. People traveled from miles around to see the rumored activities. Many came in doubt and left convinced. One young daredevil actually spent part of a night in the dark kitchen and heard the hurtling objects whizzing and banging around him.

News of "his" feats eventually reached the ears of a local medical therapist—not a doctor, as we understand the term, for he had undergone only two years of medical training—but a sort of male nurse who had the authority to treat routine illnesses and write prescriptions.

This therapist occasionally used hypnosis. He had, in addition to his other interests, a fascination for the occult. He visited the Aposhanskys and obtained permission to investigate Nina and her strange association with "him."

A word, now, about Nina. She had grown up in a comparatively well-to-do peasant family in what was then White Russia. Her parents were strict followers of the Greek Orthodox faith. In her late teens or early 20's she fell deeply in love with one Tadek, who returned her affection. There was only one hitch—he was a Roman Catholic, and his parents stoutly opposed his hoped-for marriage to Nina on religious grounds. When he turned 21 he was drafted into the army and was stationed in the Aposhanskys' city of Rovno.

Unwilling to give up her lover, Nina followed him to Rovno and obtained work as a housekeeper at the Aposhanskys'.

All went well for a time. Then the feckless Tadek underwent a change of heart. A new lass lured him away from his loyal Nina. Before long he was transferred to the town of Lubartov, about 100 miles away. Nina had absolutely no knowledge of his new address.

Nina was deeply upset. She cried often. The Aposhanskys, who had become fond of her, were at a loss. What to do?

Then "he" came on the scene, with his rappings and hurlings and that infernal dishpan.

Could there be some connection between "him" and Nina's distress? The therapist wasn't sure. He wanted to try all angles. In a series of hypnotic sessions that lasted about three months, he tried, unsuccessfully, to probe for mediumistic abilities and to regress Nina to a supposed previous incarnation—all in the interest of finding out whether she had some supernormal power.

But her supernormal powers, if any, were confined to her own immediate need to locate Tadek. The whole wild tale reached its climax one dark night when Nina asked her hypnotist to "send" her hypnotized mind to Lubartov and let her try to locate her lover.

The entire Aposhansky family was on hand for the big event. Nineteen-year-old Catherine acted as stenographer, recording everything the hypnotized girl said.

Nina went into her usual deep trance. In slow, commanding tones, the hypnotist told her, in effect, "You are in Lubartov—Lubartov—you can see the city and its buildings all around you—you can ask someone you see to help you locate Tadek . . ."

In this apparent disembodied state, Nina soon reported that she had found a man who knew where Tadek lived.

With Catherine taking notes, Nina questioned this man —or said she did—and, at the hypnotist's request, repeated everything he allegedly said to her so that Catherine could make a record of it.

The entranced Nina reported, among other things, that the Lubartov man told her Tadek's address was Kozielska Street No. 16.

Nina was then awakened.

She wrote to Tadek at the Kozielska Street address. He replied. He was, indeed, living at the precise location uncovered by her hypnotized mind. Yet Nina had never been to Lubartov. Nor had she any normal way of knowing Tadek's address.

Tadek was still associated with the "other woman." He

was unhappy, however. For her part, Nina soon found another boy friend. About six months later the mysterious "he" left the Aposhansky home, the dishpan and boots rested in peace, the milk woman resumed her deliveries, the hypnotist turned to more promising pursuits, and normalcy settled over the Aposhansky domain once again.

A wild tale from old-time peasant Poland? Catherine Aposhansky, now a happily married teacher and research technician, lived through it all and told me about it in a tape-recorded interview in February, 1969. Unlike many people who tell such stories, she impressed me as a careful observer, unlikely to be taken in by fanciful folklore.

Her story would be at least incredible, reliable though she may be, were it not for its remarkable similarity to other such accounts.

Jim Hazelwood, a newsman with the *Oakland Tribune* (California), told me of a hectic two weeks in an Oakland telephone company office at 1904 Franklin Street. During the latter part of June, 1964, employees reported a series of wild goings-on in which telephones were said to have jumped off desks, vases flew across the room, cabinets and water coolers were tipped over, bulbs broke, a dictaphone mysteriously flew out of a cabinet, and a can of coffee weighing about two pounds flew from the cupboard and made a crash landing some feet away.

An office prankster? Hazelwood stationed himself at the scene of the activity. "How," he asked me, "could a heavy metal typewriter top just take off and fly out an open window? Yet that's what happened—while I was there."

That and a dozen or so other things.

It was almost, he said, as if some mysterious energy were consciously playing pranks. He placed a metal card index file on top of a filing case and positioned himself directly in front of it, intending to keep it under close observation in the event it should be moved by the strange power. Minutes passed, and the index file refused to budge. For a split second Hazelwood took his eyes off the file and, in that instant, it crashed unceremoniously to the floor.

Clever joker? Or noisy ghost—"poltergeist?" Probably neither. Parapsychologists suspect a force from the human mind itself in these cases. Their investigations suggest that "poltergeist" phenomena may be the rampagings of a power, of which we yet know nothing, usually emanating from the mind of a young adult or teenager. Typically, so-called poltergeist occurrences take place near or in the immediate presence of some young person, and cease when the young person is absent—a suspicious enough circumstance, justifying our deepest suspicions of the involved young people. Youths like Nina. Pranksters, of course.

Take the famous "Bremen Boy" case. In a china shop in Bremen, Germany, in June, 1965, glasses and dishes unexplainably went crashing to the floor, wreaking havoc on the proprietor's patience and pocketbook. The events continued for some months, ending only when the store's 15-year-old apprentice was fired. Mysteriously, they resumed at the establishment of an electrician who hired the young man—but instead of china shattering, screws mysteriously came loose. This went on for some two years. The easy explanation would be that the youngster was playing tricks on all concerned—except for one thing. The phenomena, physically independent of the boy, were actually witnessed while they were occurring by a four-man team of German researchers, headed by Dr. Hans Bender, director of a West German parapsychological institute.

No sooner had the Bremen boy stopped his poltergeisting than another case "broke out" (there is no better term) in the city of Rosenheim, Germany. From October, 1967, to January, 1968, a Rosenheim lawyer's office was beset by weird movements of ceiling lamps and wall pictures, explosions of electric light bulbs, and the like. All these took place in the presence of a 19-year-old girl. The town fathers figured there had to be some perfectly natural explanation, of course, and the maintenance department of the town's electric company assigned its assistant director, one P. Brunner, to head up an investigation of the affair. Brunner was sufficiently impressed with what he found to report, in September, 1968, his observations to the 11th annual convention of the Parapsychological Association at West Germany's Freiburg University.

It seems he had recorded deflections of up to 50 amps on a voltage amplifier when some of the disturbances occurred—yet the house fuses, which should have blown under such a load, remained undamaged. Two physicists, Drs. F. Karger and G. Zuha, were called in. They hooked up instruments to measure the current in the main electric lines in the office. Pronounced deflections were again noted. Yet the voltage in the house mains remained constant. The experts made a careful search for faulty equipment, electrostatic charges, and fraud. Nothing of the sort was found.

When the Bremen boy and the Rosenheim girl were given psychological tests, they were found to harbor unstable personality structures with little capacity to tolerate frustration. In both youngsters, tensions from unexpressed feelings of hostility built up rapidly and were not permitted any normal outlet.

Two other cases are worth noting here. In Seaford, Long Island, and in Miami, Florida, glasses, plates, and bottles were mysteriously hurled and shattered in the presence of a 12-year-old boy (in Seaford) and a young man of 19 (in Miami).

The Seaford activity, which erupted in 1957 in the home of the James M. Hermann family, drew the attention of law enforcement officials and eventually of Dr. J. G. Pratt, who was then with the Duke University parapsychology laboratory.

When Dr. Pratt reached the scene he was met at the door by an excited Mrs. Hermann, who promptly led him to the basement rumpus room. There he found one puzzled Nassau County policeman staring in perplexed amazement at what was left of a small record player. The family's 12-year-old son James said he had been sitting at a small table under the rumpus room stairs when the record player apparently flew through the air and crashed against the lower part of the bannister.

This was typical of quite a series of crashings and smashings in the Hermann household—all happening, of course, when young James was around.

During an investigation marked by interruptions from reporters, Dr. Pratt listed a total of 67 "distinct mysterious events" which took place between February 3 and March 10. One time, for instance, Mr. Hermann was standing in the bathroom doorway, while young James was brushing his teeth and, the father maintains, he actually saw a bottle of Kaopectate move southward along the formica top of the drain a distance of about 18 inches and drop into the sink. At the same moment, Mr. Hermann said, a bottle of shampoo moved west along the same surface and slid into the floor. The formica top was level. Mr. Hermann signed a statement declaring he had actually seen both bottles start to move, apparently under their own power.

Another time, Mr. Hermann's cousin, a Miss Marie Murtha, was seated in the Hermann living room with young James and his sister when, she insists, she saw a porcelain figurine that was on a table at the south end of the sofa actually take off and sail through the air for about ten feet, landing with a loud report about two feet from the TV set. Mysteriously, it was not broken. James, she said, had been sitting in the center of the couch with his arms folded the entire time. Miss Murtha, like Mr. Hermann, attested to her observations in a signed statement.

Professional physicists and engineers tried to throw light on the matter. Studies were made of electrical influences, vibrations in the plumbing system, magnetic fields, and even the takeoff and landing times of airplanes at Mitchell Field. None of these showed any relation to the strange occurrences.

Dr. Pratt concluded that, since the phenomena were not observed under controlled laboratory conditions, it would not be proper to state that the mysterious occurrences had been caused by other than normal means. However, he said, the Seaford case added one more link in the chain of evidence suggesting that such cases are not only worth investigating, but that they deserve serious attempts by scientists to duplicate them in the laboratory.

William G. Roll, project director of the Psychical Research Foundation in Durham, investigated the Miami case in 1967. Then, aided by Dr. J. L. Artley, professor of electrical engineering at Duke University, Roll made a study of the effect of distance on the "poltergeist" activity in both the Seaford and Miami cases. The number of disturbances decreased with distance from the involved youths, the two men found, in a precisely determined mathematical ratio. Analysis of this "weakening effect" showed it fitted the exponential decay function associated with many physical processes (for example, the conversion of light to heat energy as it penetrates water). In the case of "poltergeist" activity, Roll concluded, psychical energy seems to be converted to kinetic energy.

These cases might still be written off as elaborate hoaxes were it not for their remarkable similarity to other cases throughout the world. Dr. Pratt reports that nearly 500 such cases have been given sufficient study to be recorded in serious parapsychological literature.

Actually, "poltergeist" activity ties in well with the now-famous "mind over matter" experiments which Dr. Rhine brought into world focus. In these so-called "psychokinesis" (PK) experiments, the subject tries to influence the fall of dice by "willing" them to land in a given way. To insure objectivity, machines are usually used to roll the dice.

Among the most impressive of these studies was an experiment carried out by a Swedish engineer, Haakon Forwald. Beginning in 1951, Forwald accumulated a massive amount of data showing that, by a conscious act of will, a person can guide a set of rolling cubes toward either the left or right side of a table.

Forwald *mechanically* released groups of cubes at the top of an incline. They rolled down across a table that had a line drawn down the center, dividing it into two equal right and left sides. The idea was to "will" the cubes to roll to whichever side happened to be selected as the target. Each side was designated an equal number of times, ruling out any bias in the physical equipment. The results were highly significant statistically.

Not satisfied with this, Forwald drew more lines parallel to the center one, one centimeter apart. These he numbered, forming a scale which could be used to show the actual amount of the displacement after each roll of the cubes.

It became clear, after prolonged testing, that the mind of a subject can—under some conditions at least—affect the direction of movement of inanimate objects. Forwald repeated his experiments in Dr. Rhine's laboratory at Duke University. A statistical analysis put the odds against chance at 5,000 to one. Forwald also repeated his experiments successfully in collaboration with physicist Dr. Robert A. McConnell, of the University of Pittsburgh.

The use of cubes for these experiments probably grew out of the old gambling tradition that linked hunches and "willing" with the feeling of being "hot" at the gaming tables, and with success. But there are other approaches. W. E. Cox, a researcher at the Foundation for Research on the Nature of Man in Durham, has used a variety of clock mechanisms, mercury switches, electrical relays, and the like. Cox, who has a special gift for gadgetry, has gotten significant results with these devices. In general, the idea is to try to influence electrical mechanisms—by will power alone—in such a way as to accelerate or retard clock hands, counting devices, rolling beads, and so on. In serving as a subject during brief experiments with two of these contrivances, I found that I had a definite feeling of "willing" the results during successful runs. The feeling may have been akin to the gambler's sense of being "hot." I am not prepared to state whether the impression was illusory or whether it represented something objectively real. I only know the feeling was there.

The experiments by Cox, Rhine, Forwald, McConnell, and many others have, in the minds of numerous parapsychologists, established the reality of PK beyond a reasonable doubt. PK effects may, in fact, be much more common in the lives of all of us than we realize. You may have experienced them yourself. A cherished clock stops at the precise moment its owner dies. A portrait of Mrs. Smith unexplainably falls at a time coinciding exactly with the moment her heart stops beating in a hospital 1,000 miles away. I recall that when I was in graduate school, the morning after the 1956 presidential elections, a framed copy of the bill of rights mysteriously left its place on the classroom wall and crashed to the floor. The manuscript had remained securely in place for a long period of time. There were no perceptible vibrations, no rumblings or rattlings or other disturbances to shake it loose. Was this an example of PK from the mind of a student or professor —or, perhaps, students and professors collectively? The Republican candidate, Eisenhower, had won the election handily, but sentiment on the campus at that time was running heavily in favor of Democrat Adlai Stevenson.

Other examples aren't hard to find. Dr. Louisa E. Rhine, research director at the Foundation for Research on the Nature of Man and a close co-worker with her famous husband, has analyzed a large number of such cases. Typical was the case of the Pennsylvania woman who told how her uncle, "a very strong-willed person," had designated her the executrix of his will—and how, after his death, problems arose which led her to suggest it might be wise to depart from a promise she had made to her uncle before he died.

At that point, she recalled, her husband sat straight up in bed, staring into the sitting room. "There is a chair moving across the floor!" he exclaimed. She was terrified, she said, although the partly closed door kept her from seeing into the room. The next morning, she said, she saw that the chair had indeed done a bit of traveling; the distance it had moved was marked by a curled-up rug that lay on the rung where the chair had stopped. She decided to carry out her uncle's wishes to the letter.

Sometimes it seems as though the dead use PK to let us know that something of themselves has survived the grave and passed on to another existence. A California man tells how, two nights after his wife's death, he felt her presence strongly and said, "If you are here and can hear me, give me a sign," whereupon a metal chariot with two heavy horses went crashing down from its position atop a chime clock on the mantel where it had rested securely for 12 years.

The files of parapsychologists are filled with such cases. People die—and objects fall, lights turn on and off, dishes and furniture break with explosive force, doors open and shut, locks become locked and unlocked, furniture shakes, appliances rock and rattle, bells ring, chimes sound. In one case, three clocks stopped at the exact moment of a family member's death!

Sometimes these things happen in conjunction with ESP-type experiences. An office worker tells how she suddenly, unexplainably started crying late one afternoon; about five minutes after the tears had started, a huge vase mysteriously fell from a shelf. The time was noted. The next day the woman received word that her father had died at the time the vase fell.

What are we to make of these cases? Certainly they don't prove PK, from the dying or anyone else. But if we accept the existence of PK from the laboratory evidence, it doesn't seem unreasonable to suppose that at least some of these spontaneous cases are more than coincidence.

The question is, are they really messages from beyond the grave? If Aunt Mattie's two favorite clocks stop at the exact moment she dies, is Aunt Mattie letting us know she's really with us? If a chair goes scooting across the floor when we decide to break Uncle Joe's will, is Uncle Joe registering his disapproval?

Mrs. Rhine's analysis of the cases leads her to think they're probably PK from the mind of the very person or persons experiencing the phenomena. After all, PK from the living is far better established than PK from the dead. For a scientifically minded parapsychologist, it's simpler to assume the California man himself caused that metal chariot to fall than to assume his dead wife did—given the reality of PK in the first place.

The *importance* of knowing about PK has little to do with stopped clocks and cracked vases, however. PK has far-reaching implications about who and what you are. If PK is a genuine phenomenon, it can only mean that your mind has a force of its own. Perhaps more than any other psychical manifestation, PK suggests that there's something objectively real about mind and will.

The notion of "will power" was long ago relegated to

the scientific dungeons. There was simply no place for it in the scientist's scheme of things, and there still isn't. Everything about our bodies, about the way we function, about the way we behave, can seemingly be explained quite neatly without it. Any invoking of "will power" to account for human behavior must seem, to most scientists, like a step back toward Camelot.

Now comes this reprehensible fellow, PK, threatening the bastions. The safest thing to do, in the minds of those who feel most threatened, is to ignore him. And if he won't go away, the safest thing is to try to dispose of him. You can dispose of him—or try to—by crying "fraud," by calling for sterner methods of recording the roll of the cubes, by questioning the statistics, or by questioning the experimental conditions.

In some cases, such criticisms have been justified. Where they were appropriate, improvements in the scientific procedures have been instituted. Photographic records were made of the fall of the cubes, for example, in response to one group of critics. The trouble is, the work continued to be ignored by the general scientific community even after the suggested changes had been made.

PK could revolutionize our understanding of ourselves. However you explain PK—whether by action-at-a-distance, a higher space in which our ordinary physical universe is embedded, field theories, or what—the fact remains that something sometimes happens when some people make a conscious effort to "will" a moving object without touching it. The simple act of volition, in other words, can produce measurable results. It is difficult to look at the work of Forwald and McConnell and not be impressed by this. The facts don't fit any current theories about man at all, but there they are. They cry out for a change in the way we look at ourselves—or, at least, in the way science looks at us. A fuller understanding of these events could remove us from the sterile realm of the "nothing but" man, restoring our dignity as effective volitional agents in our own right.

* VIII *

Mind over Life

Mrs. E. R., an elderly Connecticut woman, had been unable to write during the 15 years since she had suffered a slight stroke. She lacked coordination in walking and felt consistently tired. Doctors were unable to restore her lost abilities.

Desperate, she paid a visit to Oskar Estebany of Montreal, an aging former Hungarian army colonel whose large, seamed hands had apparently wrought cures in hundreds of stubborn cases.

She felt the heat—maybe that's what it was—that appeared to emanate from his hands as he passed them repeatedly over her body during a series of nine sessions in the late summer of 1966. At the end of that time she was able to walk normally, felt stronger, and—wonder of wonders—although she had not been able to write a single letter when she started treatment, she was able to write legibly after the sessions were completed.

Mrs. L. H., a middle-aged Syracuse, N.Y., woman, had suffered perpetual dizziness after being hurled against a wall on shipboard during a storm. The accident had apparently affected the semicircular canals in her ear, which govern the sense of balance. Six months elapsed before she was able to walk freely about the streets. But a number of doctors, including five ear specialists, were unable to bring her sufficient relief to allow her to return to her demanding job aboard ship.

"At first," a written record of Mrs. H's subsequent experience with the healer says, "Mr. Estebany felt that it would take three treatments to give her relief. Suddenly, after a few minutes of no sensation, Mrs. H said she was aware of heat. At first external to her ear, then within the ear. Mr.

Estebany was aware of the active response. He said she would be clear after the first treatment. A few hours later she had tried the positions that would induce dizziness and she was completely free [of it]. She took three treatments and went home feeling completely well for the first time in three years."

At a checkup two months later she reported that she had returned to her job and, instead of working a normal eight-hour day, had labored 12 to 14 hours at a stretch. The added burden had been too much, the record says, and "in the fatigued state [she] felt some of the dizziness." The record was kept by an interested physician during an observation period in Rye, New York.

Of all the weird, hard-to-believe kinds of phenomena parapsychologists have studied, I find unorthodox healing one of the most difficult to accept. Healers are exceedingly hard to study scientifically, even though they can be brought into the laboratory and asked to repeat their treatment procedures, using actual patients, under controlled conditions. The main problem, of course, is that almost every disease can improve spontaneously. There are very few conditions, short of actual amputations of limbs, that haven't—in some patients—unexplainably improved without any treatment at all. We know so little about how the normal healing process itself works that we can't say with certainty what constitutes "normal" healing and what doesn't.

Then there's suggestion. Who knows how much a patient can be helped by the mere *idea* of being helped? Every doctor has files with records of patients who "got well" after being given a prescription of doubtful pharmaceutical value. Suggestion? Yet the effects of suggestion itself are often so mysterious that perhaps it should be ranked with parapsychological phenomena—after all, you really don't "explain" anything simply by attributing it to suggestion: you merely replace one mystery with another.

Oskar Estebany is unique among healers in that his treatments have inspired some of the most careful and convincing scientific research ever done on the unorthodox healing process. But first, a word about the man himself.

He was born in Sopron, Hungary, on April 13, in 1897, of Roman Catholic parents. As a child he was apparently unusually healthy. He has taken little medicine of any kind throughout his life.

He was only an average student—a fact which stood in the way of his becoming a doctor or priest, either of which his mother wished for him. He sought a career in the Hungarian army instead. At the age of 14 he became a cadet, advancing to the rank of lieutenant during World War I. He quit the army at the age of 26, rejoining it in 1930 at the age of 33.

The first indication of his unusual healing abilities came when, while he was a cavalry officer in the mid-1930's, he noticed that animals he massaged seemed to respond better to his ministrations than they did to conventional treatments by other people. Sick animals, he found, appeared to turn naturally to him for help.

Estebany made his initial serious efforts to heal humans in 1940. Seven years later he received his first medical recognition when he treated the physician son of one of his patients—a Dr. Babits Antal, on the staff of a urology clinic in Budapest.

"Despite the fact that spiritual healing was forbidden in Hungary," Estebany told me, "I was working without any interruption, but I always worked under the observation of doctors. About spiritual healing, I am of the opinion that this is not in competition with medical doctors. It does not cure all illnesses, but it can work miracles."

The kindly, rugged-featured old soldier now lives in Montreal. He occasionally submits to tests by interested doctors and parapsychologists. One secret of his seemingly remarkable powers apparently involves his hands, which he places directly over the affected part of the patient's body.

If the procedure is succeeding, Estebany says, the patient usually feels heat emanating from those big hands—although on occasion patients have reported feeling sensations of cold and of prickliness. If the sensations are intense, Estebany claims, he feels them himself.

"From there [the feeling of heat] one can go toward discovering the illness," he explained. "But it isn't important what the Latin name of the illness is. The healing starts when one feels the root of the sickness with so much energy that this energy enables the organism of the patient to heal itself. This happens very seldom with one treatment, more often with many." There are no prayers, no religious services, no theology.

Estebany repeatedly states that he does not consider his procedures a substitute for conventional medicine. He tries to confine his treatments to disorders that have defied the efforts of regular physicians. As he expressed it to me:

"My motto is, go first to your doctor." Good advice for anyone considering the services of an unorthodox healer!

A small handful of doctors have taken a serious interest in Estebany's work. One of these, Dr. James Stephenson, a 50-year-old New York general practitioner, made a brief study of the effectiveness of Estebany's treatments in 1965. Of 20 patients who received Estebany's "laying-on-of-hands," 15 reported relief, according to Dr. Stephenson. The patients' disorders ranged from "trivial illnesses to ones generally considered terminal," Dr. Stephenson reported. Among the actual physical improvements noted by Dr. Stephenson during the three-week period of treatment: disappearance of wheezes in an asthma victim, clearing of albumin in urine, decrease in the size of a diffuse thyroiditis, increased weight gain in a severely emaciated patient with digestive disturbances, improved ambulation of a leukemic patient, and improved intercostal breathing in a post-polio patient.

Any of these conditions could, of course, have improved spontaneously. It's difficult to say what, if any, influence healer Estebany had. Maybe a lot—maybe none.

That's why the experiments of McGill University's Dr. Bernard Grad are all-important. Dr. Grad, a biologist associated with the psychiatry department at McGill, had been impressed with the possibility that some unique influence from certain human beings might alter the healing process. But, being a careful scientist, he was unwilling to accept reports of such healings simply at face value. The question was: how to rule out both that ever-present experimenter's nightmare, suggestion, and that other scientific nuisance, spontaneous remission.

Mice seemed to be the answer. Mice don't have any way of knowing that Oskar Estebany is supposed to be a healer. Thus they are immune to suggestion—assuming, of course, that they are given the same treatment by a non-healer for purposes of comparison.

If, then, you were to produce identical wounds in three groups of mice, letting Oskar Estebany treat one, letting a non-Estebany type treat another, and letting the third group go untreated, you could compare the rate of healing in all three groups.

But wait. It's a well-known fact that any fondling of animals is likely to have a favorable effect on their general physical condition. Careful experiments have shown that mice gain weight faster and grow larger if they're handled regularly. Dr. Grad got around this by keeping the mice in cages, having Estebany handle only the cages—not the mice.

In preliminary experiments carried out over a two-year period, Dr. Grad found that wounded mice healed faster under Estebany's treatment than a group of similarly wounded, untreated mice—or mice that were treated by a non-healer.

Then Dr. Grad made the experiment even more rigorous. He put the cages in heavy paper bags. One group of bags was sealed; Estebany had to hold them by the outside. The other group was left open at one end, and Estebany was allowed to reach into these bags and hold the cages—but his hands were never allowed to touch the mice directly.

Twice daily, healer Estebany held the caged creatures between his hands for periods of 15 minutes each, five hours apart. And twice daily, other people—normal, "ungifted" people—did the same thing with other caged mice for comparison. A third group of mice wasn't treated at all.

The wounds that were being treated had been produced by removing sections of skin from each mouse's back. Each wound was carefully measured by placing a thin sheet of transparent plastic over it, tracing its outline with a grease pencil, and measuring the area on the plastic.

Let's look first at the effect the treatments had on the mice whose covering bags were open. After 16 days, the wounds on the mice Estebany treated in this group were only half the size of those treated by the non-healers! There was no significant difference between the wounds on the untreated animals and those on the mice that had been treated by the ungifted people.

However, the mice whose cages were in closed bags didn't respond nearly as much to the Estebany treatment. A statistical analysis showed that the results for the closed-bag group did not approach the level of significance achieved by the open-bag group.

Does this mean that a physical healing force emanates from the hands of a few gifted people—the world's Oskar Estebanys—and that this force was partly blocked by the paper bag? If so, perhaps the force is nothing more mysterious than heat from the healer's hands. Yet the heat from other people's hands didn't have this healing action on the animals, and, further, exposing the wounded creatures to artificial sources of heat had no effect.

Perhaps the mere fact of feeling a paper bag between

himself and the caged mice had a psychologically inhibiting effect on Estebany himself, and on whatever healing powers he commands.

Whatever the explanation, the facts were there—stubborn, defiant, and seemingly inexplicable. A human being, under carefully controlled conditions, had demonstrated an unusual ability to heal living creatures just by having his hands near them. He had done this to an extent that other human beings had been unable to duplicate.

This makes no sense, in the light of what we know of biology, physics, and chemistry. But what follows is stranger still. In another series of experiments Dr. Grad compared the growth of potted plants which had received the benefits of Estebany's influence with plants that hadn't.

Dr. Grad chose barley plants because they grow rapidly and stand up straight, making them easy to measure. Now, the remarkable thing about this project was that Estebany didn't treat the plants directly at all. His "force," if that's the right word, wasn't "aimed" at the plants as such. It was directed at the solution that watered the plants!

The plants were watered with a one-per-cent solution of sterile salt water. (The salt was put in to make the plants mildly "ill," as it were, heightening the significance of any healing effect that might be noted.) Estebany held the jars of salt water for 30 minutes. Another group of plants was given water which was identical in every way except that it hadn't been held by Estebany.

Dr. Grad carried out repeated experiments with the barley seeds and salt water. In some he left the jars open. In others the jars were tightly sealed, insuring the solution's continued sterility. And in a few experiments he had mentally depressed patients hold the jars, comparing their influence with that of a psychiatrically normal person.

Consistently, the barley that got the psychically treated salt water did significantly better than the rest. In some cases it grew taller. In others, more seedlings appeared above the soil's surface and the mean yield of the plants was greater.

How common is this apparent power to heal? Do we all have it—at least in small amounts—or is it confined to a few gifted individuals? We can get a hint, perhaps, from Dr. Grad's experiment with the depressed and normal subjects. The two depressed patients apparently had an inhibiting effect on the plants' growth, Dr. Grad found—suggesting our own emotions may figure in whatever healing powers we exert.

But Dr. Grad is cautious. Simply because a person is able to influence water in such a way as to stimulate plant growth doesn't mean he has a gift for healing, Dr. Grad notes. The "power" may reside in all of us, manifesting itself only under certain conditions—and in the lives of most of us, these conditions, whatever they are, may never be realized.

Dr. Grad isn't the only scientist studying alleged healing powers. Sister M. Justa Smith, head of the biology and chemistry departments at Buffalo's Rosary Hill College, conducted a series of controlled studies with Estebany and found an apparently significant influence on the activity of the enzyme trypsin. And in Bordeaux, France, Dr. Jean Barry—a practicing physician—tried having 10 subjects retard the growth of fungus cultured in petri dishes in a laboratory incubator, using their "PK ability," as he called it, to accomplish this. Under extremely closely controlled conditions, Dr. Barry's subjects were able to do this to a degree that would have occurred by chance only once in 1,000 times.

In Durham, Robert Brier, of the Foundation for Research on the Nature of Man, tried having subjects change the electrical conductivity of philodendron leaves through PK. The subjects were told to try to change the "activity" of the plants simply by concentrating on them. Brier had polygraph electrodes hooked up to the plants' leaves to measure electrical changes.

Brier, who has a Ph.D. in philosophy and an applied background in chemical engineering, found that almost invariably he and other independent observers were able to tell which of several philodendron plants the subjects were concentrating on just by looking at the polygraph chart. At the time I interviewed him in Durham early in 1969 he had run about 100 subjects through the experiment, with uniformly positive results.

This is all so foreign to our usual way of thinking about things that to believe it is to be mildly overwhelmed. I personally would find it far easier simply not to believe any of it. This is always the handiest way of dealing with phenomena that don't fit our preconceptions—and I must confess that none of these fit mine. Telepathy were clairvoyance I can buy, precognition is tougher, but this notion that you can influence a fungus growth by thinking about it—arrant nonsense! Yet there the data sit, waiting to be explained. And there sit Dr. Grad's data, right beside Barry's and Brier's, impishly staring biological theory squarely in the face. Highly detailed, technical reports are available to any scientist who wishes to take the trouble to study them. Perhaps some explanation will emerge which is perfectly consistent with our current views of matter, energy, and the life process. Perhaps not.

Unorthodox healing fits in very closely, of course, with PK. Mind over matter; mind over life. Perhaps the complex molecular structure of living things is more responsive to PK influences than the relatively simple structure of, say, dice.

The mere act of believing seems to play a big part in it. Some people call this "faith." Faith in yourself. Faith that you can make those cubes roll whichever way you choose. Faith that there's such a thing as PK and that you can put it to work.

The mysterious healing process may be more common than we think. Dr. Emilio Servadio, former president of the Italian Psychoanalytic Society, told a London audience in 1963 that unorthodox healing influences may even play a part in the normal relationship between a physician and his patient. "It is growing increasingly clear today, I believe," he said, "that even parapsychological phenomena are part of that collective reservoir of ancestral energies upon which the individual seems to draw—in different ways and degrees—when he 'lives' the experience of his recovery."

What Dr. Servadio was talking about is really a very gray area that shades imperceptibly from faith into unorthodox healing, PK, and ESP. Nearly every doctor will acknowledge the role played by faith in the healing process. Virtually every physician would much prefer a patient who thinks he's going to get well than one who doesn't. Occasionally a doctor, having given a patient every form of treatment prescribed by science, will say something like, "We've done all we can—it's up to him now." The patient's own faith, his will to live, may then decide whether or not he recovers.

Similarly, faith—belief, if you prefer—apparently affects our performance on ESP and PK tests. Dr. Schmeidler, The City College (of the City University of New York) psychologist who pioneered the search for an ESP-Rorschach link, found that people who believe in ESP generally get higher scores on ESP tests than those who don't. Indeed, she found that when she recorded her students' attitudes toward ESP she was able—from those attitudes alone—to divide her class into high-scoring and low-scoring subjects. Her findings have been borne out time and again by other researchers. Not only that: the beliefs of the experimenters influence their subjects' ESP and PK performances! Scientists who are skeptical of ESP often get scores from their subjects that are significantly *below* what would be expected by chance—they get negative scores to an extent that can't be explained by chance alone.

Perhaps belief is a sort of bridge between conventional medicine and unorthodox healing. We have reasonably good evidence that faith plays a part in the "normal" recovery of a patient. Indeed, the role of belief is so important in the normal treatment process that scientists go to great lengths to rule out belief in their studies of new drugs. Given ESP and PK, then, isn't it possible that the unorthodox healer's own beliefs—his own faith—may play a key part in the effects of his "laying-on-of-hands," his praying, or whatever he does? Is faith really a potent force in itself?

There's a vital distinction to be made between "faith" as I am using the term here and "faith" as it is often conceived in certain religious contexts. No responsible parapsychologist has proposed that faith be used as a substitute for reason. Recognizing the power of faith is very different from blind belief in a creed, unsupported by facts. Faith, as I have used the term here, is the power of belief to change our lives, our bodies, and the lives and bodies of others. If it is genuine, it absolutely defies the narrow constructs of those who pay homage to the "nothing but" man.

* IX *

Parapsychology's Far-out Frontiers

A great many strange, seemingly inexplicable psychical occurrences are almost impossible to label. If you think you see a ghost, is it some sort of clairvoyant impression, or are you just hallucinating, or are you a liar? What about mediums? What about cases of supposed reincarnation? How about déjà vu—that strange, compelling sensation that you've "lived it all before?" What about the feeling of being out of your body?

These are the far-out frontiers of parapsychology. Parapsychologists themselves don't agree on how they should be classified or even whether they should be taken seriously at all. Yet each of these strange subjects has received serious consideration from a few researchers.

Most people in our society who think they've seen "ghosts"—or, better, apparitions—are understandably reluctant to admit it. Probably the only thing that would get you a crackpot reputation faster than seeing an apparition is the belief that you're the reincarnation of some famous person. Frank and Nora (Chapter I) thought they saw an apparition of an old lady rocking; when their children mentioned this in school, Frank and Nora began receiving irate calls from distressed parents. They soon stopped talking about the "ghost" and told their children to do likewise. In a scientific age, it seems, one is well advised not to discuss matters that don't fit the scientific picture.

It's not surprising, then, that most of our best-investigated cases come from eras and societies that are less hostile to this kind of thing than our own. Typical of these "ghost stories" is one reported by a Mr. John E. Husbands of Grimsby, England, and investigated and recorded by the British Society for Psychical Research:

"The facts are simply these. I was sleeping in a hotel in Madeira in January, 1885. It was a bright moonlight night. The windows were open and the blinds up. I felt someone was in my room. On opening my eyes, I saw a young fellow about 25, dressed in flannels, standing at the side of my bed and pointing with the first finger of his right hand to the place I was lying. I lay for some seconds to convince myself of someone being really there. I then sat up and looked at him. I saw his features so plainly that I recognized them in a photograph which was shown me some days later. I asked him what he wanted; he did not speak, but his eyes and hands seemed to tell me I was in his place. As he did not answer, I struck out at him with my fist as I sat up, but did not reach him, and as I was going to spring out of bed he slowly vanished through the door, which was shut, keeping his eyes upon me all the time.

"Upon inquiry I found that the young fellow who appeared to me died in that room I was occupying. . . ."

A letter from the hotel proprietor to the psychic investigators confirmed Mr. Husbands' story. What makes it all the more striking is that, some days later, the proprietor casually got out a picture of the dead man in Mr. Husbands' presence; without having been told who it was, Husbands identified the man in the photograph as the one he had seen.

In the lexicon of psychical research, a ghost is a special kind of apparition: the word "ghost" refers specifically to a figure that repeatedly appears in one given area—usually a house, sometimes one certain room or rooms. As the word "apparition" is used more generally, it can refer to an image of a person who is still very much alive. In the Rhines' collection in Durham, there's an account of a young woman who woke up late one night in January, 1952, to "see" her husband—who was actually miles away—standing in the bedroom. Several days later he called to say that he had pneumonia. The young man, who was undergoing Marine Corps training at Parris Island, S.C., had been compelled to stand guard duty despite a 104-degree fever the night his wife "saw" him.

In my own family there's a story about a friend's son whose face mysteriously appeared on the dining room wall while he was serving overseas in World War I. The image was allegedly seen by several members of his family. When he returned, he told them that he had been plagued by hunger at that time and, for hours on end, thought longingly of his dining room at home.

I have never seen an apparition—"ghost" or otherwise—myself. I never expect to see one. Like most 20th-century Americans, I find the whole idea hard to buy. It seems easy to listen open-mindedly while other people relate their experiences, but when it comes to actually believing in ghosts—well, all I can say is, years of conditioning prevent it. Yet hundreds of well-attested cases stand stubbornly awaiting explanation. The very first major work undertaken by Britain's Society for Psychical Research after its formation in 1882 was a census of hallucinations. The report, compiled by intelligent, critical men, shows that, of some 1,087 apparitions, 95 were seen by more than one person at the moment of their appearance. When you consider that there were only 283 cases where more than one person was even present at the time the apparition showed up, that number 95 looks pretty good. It indicates that, if you see an apparition of Uncle Joe, the chances may be better than one in three that anyone who's with you will see it too!

Is it all just verbal suggestion from one person to another, just someone saying, "Look, it's Uncle Joe," and turning pale, and stimulating the imaginations of everyone else present? But then how do we explain Frank and Nora's old lady? Frank and Nora, you remember, wrote down their—independent—impressions separately.

Then there are haunted houses. Haunted houses are strictly for Hallowe'en parties. No one believes in them. No one, that is, except people who've lived in them. Or visited them when the "ghost" or "ghosts" chose to put in an appearance. And in those cases the "ghosts" can easily be explained in terms of overwrought imaginations, fraud, suggestion, sleepwalking butlers, and so on.

Or can they?

It takes a special kind of courage to investigate a haunted house—the courage to risk the ridicule of "sensible" people. But that's the kind of courage Dr. Gertrude Schmeidler has.

In one set of experiments, Dr. Schmeidler asked nine psychically sensitive people to roam through a house thought to be haunted by a man whom the owner of the house described as 45 or older, meek, gentle, and anxious. On a floor plan of the house each sensitive marked off the spots he thought the ghost most likely to appear in. Another floor plan, marked separately by the house's occupants, showed where the apparition actually had been seen most often.

They also checked off, on a list of 57 personality traits, which characteristics they thought the apparition had.

Two of the sensitives agreed closely with the occupants on the ghost's favorite haunts. Four other sensitives' personality descriptions were strikingly similar to descriptions given by the occupants. A statistical analysis gave overwhelming odds against the results being chance.

To what, then, can the results be attributed? Strict precautions had been taken to keep the sensitives from meeting or talking with the occupants of the house or with each other. Each sensitive toured the house alone. In every case, Dr. Schmeidler paid an advance visit to the house to be certain that none of the residents were home and that no clues had been left, no lights left on, no doors left open in suggestive locations, nothing of that sort.

There remains the possibility—fully considered by Dr.

Schmeidler and never conclusively ruled out—that tradition and folklore might suggest characteristic places for a ghost to appear. A darkish hallway, for example, might be a more traditionally ghostly spot than the center of a well-lighted living room. But the fact remains that something other than chance was at work, and no one knows just what.

Then there's the home of a psychiatrist of my acquaintance—a modern, well-lighted ranch house, not at all the sort of place you'd choose to go ghost hunting. Shortly after the doctor moved in, during the late 1950's, a friend who had come to visit one evening insisted that she saw the apparition of a boy in the living room, aged about 10 or 12, dressed in the uniform of a Civil War drummer lad.

Another friend saw it a year or so later. She hadn't been told of the earlier visitor's report.

Then a third visitor—a man—saw it.

And, when the doctor married, his wife saw it.

None of these people knew that the others had seen this apparition, the doctor told me.

They all agreed on the boy's age—so, since the observations extended over a five- or six-year period, whatever he was he obviously wasn't getting any older! It wasn't as if the witnesses had actually seen a live neighborhood boy wandering in and out.

Furthermore, they all agreed that he was in uniform—although they didn't agree on the uniform's nature. One, for instance, saw him dressed in blue with a red cape and baggy, Dutch-type trousers.

Were they all having an odd hallucination which, just by coincidence, was nearly identical for each of them? Or was something else at work?

Apparitions seem to have a special predilection for people who are in their last moments of life. Every doctor, nurse, or minister who has watched large numbers of patients die has seen the look of recognition that crosses the faces of a few as they die. Typically when this happens the patient utters the name of a deceased friend or relative —almost as if the friend had come to welcome him to the "other side." Sometimes the greeting is accompanied by a few phrases which seem to suggest that the patient is communicating with a spectral visitor.

Now the obvious explanation would be that the patients are either delirious, have been given mind-influencing drugs, or are suffering from hallucination-causing disorders.

Nothing of the kind. Dr. Karlis Osis, head of research for the American Society for Psychical Research, sent detailed questionnaires to thousands of doctors and nurses asking for information about patients who had had these deathbed visions. He inquired about the patients' disorders, medications, fever at the time of death, toxic conditions, and whether, in the observer's opinion, the patient was clear-minded or cloudy-minded when death came. Dr. Osis followed these up with telephone interviews.

In the course of two such studies, Dr. Osis got replies from a total of 1,645 doctors and nurses. Surprisingly, patients who "saw" dead relatives and friends tended to be clear-minded, not on drugs, free of fever and toxicity, and aware of their circumstances and surroundings. By contrast, patients who hallucinated angels, devils, monsters, or living people did tend to be on drugs, cloudy-minded, or suffering from hallucination-causing disorders.

Surprisingly, too, patients who "saw" dead relatives and friends did not recover after the vision. From the comments made by these patients, it was clear that, in many cases, the apparition's visit was directly related to the patient's death and transition to what the patient believed was another life.

But when a patient—usually cloudy-minded—saw a hallucination of a person who was living, the hallucination never had anything to do with a "next life."

When the apparition's visit was aimed toward helping the patient over the transition called death, the observers usually noted a marked change in the patient's attitude toward dying. In most such cases, patients became serene, peaceful, and content.

A happy, comforting thought. Few parapsychologists would assume, of course, that this or any other visionary or hallucinatory experience proves that there's a life beyond the grave. Such cases do offer a basis of hope for those who wish to believe. But a simpler explanation, in the minds of most careful parapsychologists, is that apparitions are a sort of dramatized ESP. We have seen how this can happen in the more conventional cases of telepathy and clairvoyance (Chapter IV).

But that, of course, is only a theory. For all anyone knows, an apparition of a dead person may actually be a sensory dramatization of a telepathic message from the dead person himself. It may be much as if the dead person were somehow sending instructions to the mind of the living percipient, providing the living person's sensory apparatus with a sort of blueprint for creating a sensory image of the deceased's physical appearance.

Actually, there are much more mystifying apparitions than the conventional, garden-variety figures of dead people. Most puzzling of all, in my opinion, are the apparitions of living people who, at the very time their images were seen, were actually experiencing being "out of" their own bodies and were accurately observing events at the locations where their apparitions appeared. I suppose my own partial out-of-the-body experience, which I described at the beginning of this book, fits this category. But there are far better examples. Here, in shortened form, is a classic case reported by Mr. and Mrs. S. R. Wilmot and investigated by the British Society for Psychical Research (Mr. Wilmot is the writer):

> I sailed from Limerick to New York on the steamer *City of Limerick*. On the second day out a severe storm began. Upon the eighth day the tempest moderated and for the first time I saw my wife, whom I had left in the United States, come to the door of the stateroom, clad in her nightdress. At the door she seemed to discover that I was not the only occupant, hesitated a little, then advanced to my side, stooped down and kissed me.
>
> Upon waking I was surprised to see my fellow traveler, whose berth was above mine, but not directly over it—owing to the fact that our room was at the stern of the ship—looking fixedly at me. "You're a pretty fellow," he said, "to have a lady come and visit you in this way." I pressed him for an explanation; he related what he had seen while lying wide awake in his berth. It exactly corresponded with my dream. He was a sedate and religious man whose testimony upon any subject would be taken unhesitatingly.
>
> The day after landing I went to Watertown, Connecticut, where my wife and children had been for some time. Almost her first question was, "Did you receive a visit from me a week ago Tuesday?" "A visit from you?" I said. "We were more than a thousand miles at sea." "I know it, but it seemed to me I visited you." "It would be impossible. Tell me what makes you think so."
>
> My wife then told me that on account of the weather she had been extremely anxious about me.

She had lain awake thinking of me and at about 4 a.m. it seemed to her that she went to see me. "Tell me," she said, "do they ever have staterooms like the one I saw, where the upper berth extends farther back than the under one? A man was in the upper berth, looking right at me, and for the moment I was afraid to go in, but soon I went up to the side of your berth, bent down and kissed you."

The description given by my wife of the steamship was correct in all particulars, though she had never seen it.

More people have had these out-of-the-body experiences than care to admit to them. Just describing one is almost as sure a confession of psychosis, in the minds of many people, as telling of seeing a ghost. Yet I never cease to be amazed when, on bringing the subject up among trusting friends, I not infrequently find at least one person in a group of a half-dozen who will admit to having had such an experience. They are sane, sober people, and they uniformly insist they were not asleep at the time.

The Rev. Calvin G. Sheasley, Pittsburgh Methodist minister, told me about a series of such experiences he had, beginning in September, 1968. It started while he was meditating in an easy chair in his living room shortly before retiring late one night. Suddenly, he said, he felt himself rising. He was aware of his body on the chair below him.

"It was like being in two places at once," he recalled. "I knew that was my body down there and that I was identical to it, yet I also knew I was separate from it and could look back on it.

"I saw myself approaching the ceiling. As I got near it I was afraid I'd hit it. As soon as I felt that fear—Bang!—I was back in my body."

If such a thing ever happened again, he resolved, he definitely would not succumb to fear—he would, in effect, "let himself go" and see what happened.

He got his chance. One night in October, 1968, some time between midnight and 1 a.m., he was sitting in that same chair meditating before going to bed. Again he felt that rising sensation and that separateness from his body.

He reached the ceiling. Then, with no further sensation of traveling, he found himself standing in a doorway leading to a friend's living room in suburban Baltimore, over 200 miles away. This friend—Mrs. Idella Watts, a retired school teacher—had lived alone for many years. Mr. Sheasley had visited the house and was familiar with the general setting and with the interior arrangement.

"As I stood in the doorway," he said, "I saw her sitting in a red chair that she had always kept in front of a book case, between the dining room and living room. But this time the chair was in a corner. I saw her sleeping there with her mouth open. I heard footsteps down the hall and wondered who it was, since she lives alone and doesn't usually have visitors. I looked over and saw that her cuckoo clock, which she always kept running, had stopped."

The whole experience, he said, was a bit frightening. After all, he had never done this before and he had no assurance that he could ever get back.

But the moment the thought of returning occurred to him, he said, he found himself at home in his chair again.

The following night he received a phone call from Mrs. Watts. Her call was apparently a response to one of those irrational impulses that prompt us all, at times, to do things we can't explain. Perhaps, as she herself hinted, she had sensed Mr. Sheasley's presence the night before and was responding to that.

She confirmed that she had been sitting in the big red chair and had fallen asleep at the time he "saw" her. She also confirmed the new position of the chair, the stopped clock, and the fact that she had a visitor—a young woman —staying overnight.

Mrs. Watts verified most of the details of this experience in a telephone interview with me. She said she had had a strong impression that Mr. Sheasley had had a message for her before she called him. "I had this urge, like it was just saying, 'Call Cal, call Cal,'" she declared.

Mr. Sheasley had another dramatic out-of-the-body experience two months later. He was sitting in his living room one Friday night shortly before Christmas. Being in considerable distress following an illness, he had decided to go to bed early.

"I suddenly felt a coldness, a kind of numbness in my body—like when your leg or arm goes to sleep, only all over," he said. "At that moment I had this feeling of leaving my body—of separation. When I reached the ceiling this time, the next thing I knew I saw a mountain, a stream, and a waterfall. I was conscious of coming on it high above, but it wasn't a sensation of flying. It was something like looking at a Cinerama movie from the balcony.

"Then I saw a clearing and a cabin. It was kind of odd

because one side was concrete block and the other was wood. It wasn't an old cabin. I didn't go inside, and yet I knew a minister friend of mine was there. I sensed that others I knew were there and I had a feeling of remorse that I couldn't be there. As soon as I felt that remorse I was back in my chair."

Mr. Sheasley said he had known the other minister was out of town but did not know the man was visiting his cabin in the Allegheny Mountains. Nor had Mr. Sheasley ever been to this cabin.

When he encountered his friend in town a few days later, Mr. Sheasley asked him:

"Did you have a good time Friday night?"

Yes, his friend said, he had.

"Why don't you invite all your friends to your parties?" Mr. Sheasley inquired.

His friend, who is skeptical about things psychical, replied in astonishment:

"How did you know that?"

Mr. Sheasley then described the cabin including such details as a concrete step in front of a plank door, a bush at the left of the step, and vertical paneling on one side of the house. His friend corroborated his description in all particulars.

Mr. Sheasley's out-of-the-body experiences are not unlike hundreds of other such experiences that have been reported, collected, and compared over the years. Common features in these cases include seeing one's own body, a feeling of rising, the feeling of heading toward the ceiling, and the sense of viewing events from a definite point in space—as if the observer actually occupied a position in space himself. Many people also report an impression of traveling to and from distant points in space.

People who have had these out-of-the-body experiences insist that they aren't dreams. Even patients whose out-of-the-body experiences occurred during deep comas stoutly maintain they weren't dreaming. A Montana school teacher told me how, during a serious illness, she experienced the sensation of leaving her body and of observing her doctor and her husband discussing her illness. She was in a deep coma and wasn't expected to live through the night. She seemed to watch, she said, while her husband and the doctor went into another room, removed a book from the book case, and referred to a specific passage in it. After she recovered, she said, she told her husband of the experience and he confirmed having removed that precise book at that time.

Are people who have these experiences deluded, or lying, or just plain crazy? There's only one sure way to find out, of course, and that's to study and observe them under controlled laboratory conditions while they're having these out-of-body projections. And that's exactly what Dr. Charles Tart did.

Dr. Tart, of the psychology department at the University of California in Davis, was fortunate enough to find a subject who claimed to have had these out-of-the-body experiences two to four times a week throughout much of the 20-odd years of her life.

Dr. Tart calls her Miss Z. He describes her as a warm, highly intelligent young woman. At the time of the experiment she had completed two years of college and had interrupted her education to earn enough money to continue at college.

During a typical out-of-body experience Miss Z would wake at night to find herself floating near the ceiling. She often saw her body lying on the bed during the brief moments the experience lasted. After about a half-minute she would go back to sleep. Usually that's all there was to it. She never thought much about it—during childhood she just assumed that everybody had these experiences; but when she mentioned them to friends during her teens, she discovered, to her surprise, that they weren't so ordinary after all. Her friends, she found, thought they were definitely on the odd side. So, of course, she stopped mentioning them. But of one thing she felt certain—they weren't dreams.

Only on one or two occasions did she ever experience going beyond that ceiling. One occurred when she was about 14—and it apparently resembled a vivid nightmare, so it may not have been a typical out-of-body experience. Anyway, in this state she found herself walking down a dark street in her home town. It was a deserted district. She had on a checkered skirt, she noted, which was unlike any she owned. She had the distinct feeling that she was "in" someone else's body. A mysterious figure was behind her, following her. She was terrified. He seized her, raped her, and stabbed her to death. She woke in horror. The "dream" had seemed frightfully real.

The next day, she said, a local newspaper carried a report of a girl who had been raped and stabbed to death in that exact neighborhood. The girl was described as wearing a checkered skirt.

This would probably be classified as an unusually vivid clairvoyant dream were it not for the frequency of her out-of-body projections.

Dr. Tart was impressed by Miss Z. Here, perhaps, was a rare opportunity to study a person who was actually having an out-of-body experience.

In a sort of "pilot experiment" he asked her to cut up 10 slips of paper, write the numbers from one to 10 on each, put them in a large cardboard box, and shake them up. Each night after she was in bed she was to draw out a slip of paper at random and put it on her bedside table without looking at it. She couldn't see the number from her position in bed. If she awoke during the night and found herself floating near the ceiling, she was to memorize the number and then check after she awoke to see if she had perceived it correctly.

Miss Z tried this for seven nights and told Dr. Tart she had gotten the number right each time.

Scientifically, of course, this was worthless. There were no safeguards against Miss Z inadvertently seeing the number, the process of randomizing the numbers left much to be desired, and, worst of all, she had to rely on her memory to tell her which number she had perceived during the night.

But it was helpful in that it suggested that, perhaps, a truly scientific study in a special sleep laboratory would be worth undertaking.

The laboratory was divided into two sections—a sleeping room, equipped with instruments for recording vital changes going on in the subject's brain, heart, breathing, and circulation, and an observation room where the experimenter, Dr. Tart, could sit. A large viewing window between the rooms was covered with a Venetian blind to darken the room enough so that Miss Z could sleep.

The reclining Miss Z was hooked up to lie detector equipment, a heart monitoring instrument, a miniature strain gauge to record her eye movements, and an electroencephalograph to chart the electrical impulses from her brain.

There was enough slack wire from all this to let her turn over in bed, but if she sat up more than two feet she would have pulled the wires out and all hell would have broken loose on the recording instruments. Any movements beyond the two-foot limit would have been spotted by Dr. Tart immediately.

Just above the observation window—about five and a half feet above Miss Z's wired-up head—was a five by ten inch shelf. Immediately above the shelf a large clock was mounted on the wall. Each night after Miss Z was in bed, Dr. Tart wrote down a randomly chosen five-figure number on a piece of paper. Carefully concealing it from Miss Z's vision, he placed it on the shelf above her bed. She had been instructed to try to rise high enough to read this number. If she succeeded, she was to tell Dr. Tart immediately, via an intercom system, what the number was. Dr. Tart, meanwhile, would monitor her electroencephalograph waves, heart rate, and so on. Miss Z kept him posted on every development in her out-of-body efforts. Every time she had an out-of-the-body experience—or a partial one—she told him so right afterward.

The first night was just a get-used-to-it session, with the thoroughly wired Miss Z trying to accustom herself to her unfamiliar circumstances.

The next night the phenomena began. Miss Z, who by now was used to being hooked up like an astronaut, began to behave like one. She rose up, she said—out of her body—but not quite high enough to read the target number. For what it's worth, she did read the clock—"Write down 3:15 a.m.," she mumbled as she awoke. "I don't see the number, but I just remember that." This, she said later, was merely her attempt to communicate that she had been "out," and had risen a bit.

Meanwhile, Dr. Tart's equipment was busily recording alpha waves, spindles, in-between waves, heart rhythms, eye movements, and that good old standby, the galvanic skin response.

Three nights elapsed, with Miss Z reporting varying degrees of out-of-the-bodiness. Perhaps because of the distracting influences of her unusual surroundings, she couldn't quite seem to make it up to the shelf.

Then, at the end of the fourth night, it happened. At 6:04 a.m., reports Dr. Tart in the January, 1968, issue of the *Journal of the American Society for Psychical Research,* Miss Z woke and called out that the target number was 25132. "This was correct," Dr. Tart stated, "with the digits in correct order, but I did not say anything to her at this point; I merely indicated that I had written the number down on the record."

Fraud? Had Miss Z concealed mirrors and reaching

rods in her pajamas, using them to read the number? Not impossible, but highly unlikely, Dr. Tart said. There was also a remote possibility she could have seen the target number reflected in the clock face, although there wasn't nearly enough light in the room to see the reflection with normal vision. When Dr. Tart and a colleague tried doing this themselves, they found the room far too dark; only by holding a flashlight directly over the number, increasing its brightness several hundredfold, could the men "just make out" what the number was.

"Thus," Dr. Tart concludes, "although it seems unlikely, one could argue that the number constituted a 'subliminal stimulus' in its reflection off the clock surface."

In other words—unlikely though it may seem—there's the remote possibility that the number reached Miss Z's awareness through her normal senses but at a level of sensory stimulation below that usually measured by scientists. Highly unlikely, yes—but no good scientist I've ever known will overlook even the most far-out alternative to ESP, and rightly so. Dr. Tart, apparently, is a good scientist.

More conclusive, perhaps, are Dr. Tart's observations of Miss Z's bodily changes while she was having these out-of-the-body experiences. Her heart rate didn't drop off to the slow, deathlike levels of the deeply comatose patient. It kept pounding along at 72 beats a minute. But her electrical "brain wave" tracings were unusual. Dr. William Dement, recognized among the world's foremost sleep experts, studied the tracings and agreed with Dr. Tart that they couldn't be classified into any of the known stages of sleep. Nor were they typical of waking or drowsy patterns.

And the eye movements that accompany ordinary dreams weren't present, either.

It adds up to this: Miss Z apparently wasn't dreaming, in the usual sense at least, while she was having these out-of-the-body experiences. Both her brain wave patterns and her lack of rapid eye movements testify to this. Nor was she awake, in the traditional meaning of the word.

What—or where—was she? Is this another strange, dramatized form of ESP? We simply don't know. Nor do we know how many people have these out-of-the-body experiences. We can't tell. Perhaps we all do, and simply don't remember them the next morning—just as most of us forget our dreams. Or perhaps this ability to "travel" out of the body is an off-beat sort of gift, given only to a few individuals.

Parapsychology's Far-out Frontiers

Stories of psychically gifted people have studded the folklore of the human race. Many—perhaps most—of these people have been frauds. Nevertheless, a small number apparently have shown genuine psychic ability. A few of these—chiefly mediums—have been responsibly studied.

The phony atmosphere that surrounds most seances is enough to frighten away nearly any thoughtful scholar. And understandably so. I haven't had many experiences with mediums myself; the ones I have had have left me feeling utterly foolish—filled with self-reproach at the thought of having allowed myself to be part of such blatant nonsense.

An exception was one meeting at a spiritualist camp in Freeville, N.Y., in the summer of 1961. My wife, I, and several friends were attending what promised to be the usual vaguely generalized service. Late in the evening one of the mediums looked at me and said:

"There's someone on the Other Side who has a message for you. She says her name is Helen. They called her Nellie."

I insisted that I had never known a Helen nicknamed Nellie. The medium insisted that "Helen" was there and added that the message concerned me indirectly—that I should ask other members of the family about "her."

Four days later we went to another service. They had an entirely new group of mediums in. Presumably, if there had been any exchange of information between the first group and the second—any foul play—the mediums in the second group would have been warned to stay clear of any references to "Helen" in my readings.

As it happened, however, the new medium plunged right in with the "Helen" bit as soon as she saw me. "Helen," she insisted, was there trying to get through, and perhaps I should consult another member of the family.

Puzzled, I later asked my mother during her next visit.

Without a moment's hesitation, mother told me that a Helen Fox—who was called "Nellie"—had borrowed $10,000 from my grandmother under false pretenses, although my grandmother counted her among her best friends. At least part of the money would have been mine today, she said.

It is highly unlikely that these mediums could or would have conducted any investigation into my remote family history to uncover this fact. Both my visits were spontaneous, unplanned affairs for one thing, and, since I was not a newspaperman or well known at the time, there was no motivation for them to make extensive inquiries—I was just a face in a crowd. For another thing, my mother was probably the only living person who knew of this incident.

Truly gifted mediums are rare. Careful study of the few who have been available hasn't proved that the dead can communicate. Given highly developed ESP ability, parapsychologists say, a medium could get any of the information yielded in seances without any "communication from the dead" being involved. Mrs. Eileen J. Garrett, President of the Parapsychology Foundation, and one of the most carefully studied mediums in modern times, is by no means convinced that her apparent "communications" really do emanate from departed spirits.

In short, parapsychology has really done more to challenge the "communication-from-the-dead" idea than to support it. When research showed the possibilities, not only of telepathy and clairvoyance but of precognition, it became clear that there was almost no way to rule out the possibility that these—and not departed spirits—were the agents.

None of this means, of course, that other-worldly communications (if there is another world) do not take place. Many people who have had wide experience with mediums say that they are most convinced by the subtle shades of the "departed spirit's" personality which the entranced medium projects—a characteristic gesture, perhaps, a pet name, a favorite phrase, plus a persuasive feeling—that the dead person is really there.

Like so many strange things in parapsychology's far-out frontiers, this, too, must remain unanswered until someone devises a way by which it can be scientifically tested.

But communication with departed spirits isn't the only possible approach to the question of survival after death. A more unusual approach has been taken by Dr. Ian Stevenson, former Chairman of the Psychiatry Department at the University of Virginia School of Medicine. Even to parapsychologists, Dr. Stevenson's hypothesis represents one of the farthest-out of parapsychology's far-out frontiers. Yet his early efforts won him, in 1960, the American Society for Psychical Research's William James Essay Award for research bearing on the subject of survival.

The possibility seems almost too bizarre for Western minds to accept, yet Dr. Stevenson's hypothesis is that some small children can actually remember former lives on the earth—previous incarnations.

Dr. Stevenson has impressed me as being a careful observer. His investigations have led him into virtually every corner of the earth, interviewing, checking, cross-checking, examining records.

Most spontaneous pre-birth memories, he has found, seem to appear in children under the age of four and disappear by the time the child is nine or ten.

In the most convincing of these, the child insists that he lived in another city or town, gives his alleged former name, the names of his supposed former parents, and—on being taken to his alleged former home town—is able to find his way without guidance to his supposed former house, correctly naming the occupants, giving intimate details of his alleged past life, and even identifying changes that have been made in the house since he "lived" there. In some cases, these identifications have been made in the midst of strangers who deliberately tried to mislead the child.

One such case involved a Ceylonese youngster by the name of Gnanatilleka who, at the age of one year, began referring to "other parents" she claimed to have had. By the time she was two, she was making clear references to a previous life. She had been born in Hedunawewa, in central Ceylon, yet she insisted that she had lived in another town at another time. She said, among other things, that she had had two brothers and many sisters, that her father had been a postman, that one brother's name had been Dharmadasa, and that her mother had been stout.

Although she didn't give any definite location for her previous life at first, a visit by some villagers from a town called Talawakele prompted her to state that Talawakele was her former place of residence.

She asked to visit her former parents, gave details about her former home and the names of members of her former family.

A journalist in the city of Kandy, Mr. H. S. Nissanka, heard of Gnanatilleka's claims. Aided by a Mr. Piyadassi Thera, Nissanka—using the descriptive material furnished by Gnanatilleka—was able to identify a family in Talawakele which matched the girl's description. The family had had a son, Tillekeratne, who had died Nov. 9, 1954, at the age of 12—15 months before Gnanatilleka's birth. Gnanatilleka's description of her previous life matched Tillekeratne's perfectly.

Gnanatilleka's family took the girl to visit Talawakele for the first time when she was four years old. She recognized a number of buildings in the town—although she had never been there. She guided her family to the very spot her supposed former house—Tillekeratne's home— had occupied. Alas, the house had been torn down. Shortly before Tillekeratne's death, his family had moved. The two families did not meet on Gnatatilleka's first trip. However, Gnanatilleka did meet three of the teachers from Tillekeratne's old school when they visited her in her home at Hedunawewa. She recognized them and described details and events at the school when "she" (as Tillekeratne) had been there.

Early in 1961 Gnanatilleka again visited Talawakele. This time she did have a chance to meet Tillekeratne's family. One by one, his relatives and acquaintances were brought before Gnanatilleka. The occasion was witnessed by Mr. Nissanka, Mr. Thera, and one of Tillekeratne's teachers, Mr. D. V. Sumithapala.

I talked with Mr. Nissanka about this in December, 1968, while he was in the United States studying political science at the University of Pittsburgh. He assured me that she not only identified seven members of Tillekeratne's family, and two others in the community, but that she did this while other people were actually trying to mislead her!

Equally impressive, Nissanka said, was that she failed to recognize people whom Tillekeratne had not known—although, in an effort to test her to the fullest, some had come forward and falsely identified themselves as Tillekeratne's friends.

Gnanatilleka gave intimate details of Tillekeratne's life —his father had been a postman, he went to school with one sister, another sister had attended school in a place called Nawalipitya, she had a brother called Dharmadasa, she recognized a house where the laundryman had lived, she correctly described and drew a diagram of Tillekeratne's school and the railway station near the school, including a bridge and steps between the station and the school, and gave a great many other minor details

Tillekeratne had been very close to one of his teachers, Mr. Sumithapala, and Gnanatilleka's behavior toward him was appropriate to this. Mr. Sumithapala had taught Tillekeratne a mythological tale, which Gnanatilleka was able to repeat in great detail with dramatic gestures—although she had never heard the story in her family.

Throughout the investigation, Nissanka said, the Communist press in the area maintained a barrage of ridicule.

Belief in reincarnation, it seems, was a bastion against the expansion of Communism in the area. So the Communists did all they could to discredit it—unsuccessfully.

Gnanatilleka is only one of more than 600 cases Dr. Stevenson has investigated. He has reported 20 of these in meticulous detail. Many of these are as impressive as the case of Gnanatilleka.

What are we to make of cases in which small children insist that they belonged in other families—often in other cities, where they have never been—and, on being taken to those cities, are able to locate their supposed former homes, correctly identify the inhabitants, and give details of the lives of family members whom they claim to have known?

What are we to make of cases in which these youngsters show the same strong emotions toward various members of their supposed former families as their alleged former selves showed? What are we to make of cases in which, for example, a child claims to have once been the mother of a person in another town—someone he has never met—and, on being introduced to this person, treats him exactly as his mother did?

Case after case is on record. The big debate among parapsychologists is whether reincarnation is really the best explanation, or whether some form of ESP isn't the answer. Isn't it more likely, they ask, that these children are unusually gifted psychically, and that they are able to pick up details of other people's lives clairvoyantly or telepathically?

Dr. Stevenson's reply is that these children do not show any unusual ESP ability or mediumistic powers. Further, he notes, their patterns of recall have no relationship to any known patterns of ESP development. For the most part, they seem to remember their "former lives" most clearly when they are very small—usually under the age of five—and by the time they are six or seven their memories start to fade, and they begin to lose interest in their supposed former identities. By the time they're nine or ten they've usually stopped talking about their "former selves" entirely. This looks much like the natural process of forgetting—as would be expected in a memory of a former life.

Not surprisingly, many of the best cases are found in cultures where reincarnation is generally accepted. I say this is not surprising, because children in non-reincarnationist societies would be much less likely to get a fair hearing from their parents, their parents would be less likely to report such stories to anyone else, the child's story would more likely be taken as fantasy and ignored, and the child himself would be less likely to interpret his apparent memories as reincarnation.

Yet there are such cases reported in the United States, Germany, England, and other predominantly Christian nations. They are generally harder to verify. It's more difficult to check out names, places, and events.

I have come across one such case myself. The mother, a highly intelligent school teacher, is not at all sure reincarnation is the best explanation. She has asked me to keep her identity and that of her daughter in strict confidence.

I will call the girl Suzanne. Her first reference to reincarnation came at the age of three years, shortly after the death of her great grandfather. Her mother had been explaining the old man's demise to her, using such phrases as "Grandpa wanted to see God, and God decided he'd like to see Grandpa, so Grandpa died."

Suzanne's reply: "Grandpa is coming back."

Her mother asked her what she meant.

"After you die," Suzanne said, "you go up to heaven, then you come back again as a little baby."

"You mean Grandpa is coming back again?" her mother asked.

"Yes," she said.

During the next two years Suzanne, whom her mother and grandmother describe as exceptionally advanced, made continued references to the process of dying and coming back again as a baby. She at no time used the term "reincarnation," however. Had she done so, one might suspect that she had picked up the concept from some acquaintances in the here and now.

From time to time she made such comments as, "It will be okay when Geegee (a pet name for her great grandmother) dies, too, because she will come back again."

On New Year's Eve in 1968, Suzanne's grandmother was putting her to bed, when the child mentioned that she could remember living before. She asked her grandmother not to tell her mother about it, however, because "mother would feel bad if she knew I used to have another mother."

She then described a life she said she had lived in Africa.

The grandmother did tell Suzanne's mother about the conversation, however, and her mother tactfully broached the subject to her, telling her to feel free to relate any such impressions she might have.

In subsequent discussions with her mother, Suzanne described what she said was a life she had lived not too far from the coast of Africa, in a home that was isolated from everything except one store. There were two possible routes to the store, she said—one, which her father didn't like, was marred by quicksand. She said the store carried cloth and thread, among other things, plus food—"but we don't eat food like that now."

She described a common food as being white and "fluffy," gesturing with her hands to show what it looked like. Her mother thinks she may have been referring to manioc.

Suzanne said that she did not get along well with her former mother. She liked her father better, she said, but added that he was away from home a great deal. She said that much of her former life had been sad and lonely. But when her present mother asked her to elaborate on this, she said, "I'm only going to talk about the nice things that happened." She persistently refused to talk about the unpleasant things.

Much of her comments centered on her dolls and toys. She went into considerable detail as to how the dolls were made. She told of a woman who sold clay of different colors, and mentioned a process by which these colors were created. The method of making hair for these dolls—which she apparently did herself—seems to have involved attaching white strings to the clay head.

At one point, she told her grandmother: "I didn't have crayons like this then. I used feathers." Later when a historical television show portrayed men using quill pens she commented, "Those are like the feather crayons I used to have."

As she described the clothes, they seem to have resembled hoop skirts. She repeatedly stated that "we didn't have much in Africa," and once followed this with the statement that they had used wooden bowls and cups of their own making.

Suzanne's mother says the child is truthful and is not prone to idle fantasy. But, intellectually, the mother says, Suzanne would be capable of fabricating the memories of the African "life."

Yet when she has asked Suzanne about this, the girl has flatly stated that these are genuine memories. Asked whether she ever made up details to fill in blank spaces in her memory, Suzanne acknowledged that sometimes she did try to "fill in" in this way, but that other memories of the experience were true and very clear.

Her mother still isn't entirely satisfied that reincarnation is the answer. In talking to Suzanne on the subject, her approach has been to listen and do very little commenting herself. She asks few questions, she says, because she doesn't want to strengthen the fantasy—if it is a fantasy —but at the same time she wants to keep the channel of communication open. The mother never initiates the subject herself, and when the child initiates it, her mother doesn't ply her with leading questions.

The most convincing part of the story, from her mother's standpoint, is its immediacy—Suzanne seems to be describing actual events in a factual manner.

Parapsychologists themselves are understandably cautious in their approach to reincarnation. A common objection is that we have no proof that there is a "spirit" that survives death at all, let alone one which reincarnates. The problem of survival itself should come first, they argue—then reincarnation.

Reviewing Dr. Stevenson's work in the December, 1966, issue of the *Journal of Parapsychology*, Dr. Louisa Rhine concludes:

"The material in this book [*Twenty Cases Suggestive of Reincarnation*, American Society for Psychical Research, New York, 1966] must therefore be taken as a collection of cases which by their very nature have built-in weaknesses that make their interpretation very uncertain. . . . All of these strictures together do not detract from the fact that the author has taken the trouble to make the difficult collection of cases of claimed reincarnation and has presented his material carefully and interestingly, and that he has thereby given parapsychologists and any others interested the opportunity to form their own opinions of the kind of material it is."

More common than these seeming pre-birth memories —and therefore easier to confirm—is the strong feeling of familiarity we sometimes get in places we've never been. The experience, known as déjà vu, can be disturbing, anxiety-provoking, even terrifying, depending on how well you can tolerate the unknown. It can be so persuasive, so hauntingly provocative, that the very universe itself seems, for the moment, to remain hung in eternity while past and present vie with each other. You are sure you've lived this moment before—but yet not sure. (How can such a thing be possible?) You know you haven't seen this spot, heard this conversation, thought these thoughts—and yet you know you have. You swear you could almost find your way without directions. And that's just what happened to a Pittsburgh woman.

Ruth Lawrence was visiting Oil City, Pennsylvania, for the first time, in early November, 1959. While her brother attended to a business matter in town, she stopped in at a clothing store to do some shopping.

"The moment I stepped off the elevator onto the second floor I had this overpowering feeling of having been there before," she recalled later.

"It was such a pleasant feeling—it was a feeling of warmth and security—it was the feeling, 'This is home.'"

Usually, Miss Lawrence is irritated by clothing store clerks. She feels that they often try to sell clothing to her before her mind is made up. But when she stepped off the elevator in the Oil City store, that annoyance was strangely absent.

"I was amazed that here I wasn't irritated," she said. "But all the time I was there I had this overwhelming feeling, 'Oh, I've been here before, isn't it good to be here!'"

She felt, literally, like a different person, she said—a different identity. It was almost as if all her usual feelings and associations had been stripped from her and replaced by new ones. People, objects, and events no longer had the same connotations for her. Dealing with the clerk, she said, was a pleasure. Through it all she felt strangely light and dreamlike.

Up to then the day had been rather upsetting. She had not wanted to make the trip. "It was not a day I looked forward to at all," she recalled.

Miss Lawrence was shown several dresses by the clerk, who then left her to tend to affairs in another part of the store. Still in the grip of déjà vu, Miss Lawrence went directly to an unmarked, inconspicuous dressing room without being told its location. It lay behind a plain, ivory-colored door—but it could have been behind any of a half-dozen other plain, ivory-colored doors in that plain, ivory-colored room. There was absolutely no way a stranger would have known which door led to it.

"I tried to seek a logical explanation later," Miss Lawrence said. "I asked myself if perhaps I might not have once seen a store similar to this. But it wasn't a physical similarity that prompted this feeling. It was more of a mental atmosphere."

The odds against Miss Lawrence's finding the room by chance are not overwhelming—perhaps five or ten to one—but the important thing here is Miss Lawrence's insistence that she *knew* where the room was. If you're in a strange house looking for the washroom and stumble on it by chance, you know you've found it by coincidence; but if you're in a house you've visited before, and you've been in the washroom and know where it is from your previous visit, you know that your ability to locate it is not just coincidence. Miss Lawrence apparently was able to locate the correct room because of an actual feeling of *knowing* its location. But she had no such familiar feelings about the town of Oil City itself—just the store.

She is, she said, absolutely sure she had never visited Oil City.

"Even as I left the store," she recalled, "I was trying to get my bearings."

This "other identity" feeling persisted throughout the drive home. It was so strong that, for the first and probably last time in her life, Miss Lawrence had no desire to return to her family. This was totally unlike her. "It was an utterly foreign feeling to me," she declared.

Miss Lawrence has no interest in reincarnation and did not consider it as a possible explanation.

A number of theories have been advanced to explain déjà vu. Some psychologists think it's simply a response to a scene that has strong similarities to some other scene or event in the observer's life. Another theory is that part of the brain momentarily races ahead of the rest, perceiving events and circumstances before the rest of the brain. Neither of these explanations would explain how Miss Lawrence was able to find the dressing room without guidance. A third theory, held by many parapsychologists, is that déjà vu is a vague recollection of a precognitive dream—a memory of a dream one had about the event before it took place.

None of these theories has been proved. Lacking concrete evidence, déjà vu must remain in the limbo of the unexplained, along with apparitions, out-of-the-body experiences, mediums, and seeming memories of former lives. These are all part of parapsychology's far-out territory—land that is uncharted and controversial even on the parapsychologists' dimly bounded maps. Some would say it's still too soon to try to explore it meaningfully. Others would say parapsychology has already advanced beyond it.

Whatever we decide, these theories will remain—tantalizing, haunting, challenging, like the shreds of a dream one can't quite grasp before dawn dissolves it in wakefulness.

* X *
Will Russia Score First?

In October, 1957, Soviet scientists shocked the world—and, most particularly, the United States public—by launching mankind's first orbiting spacecraft. There is little question that, had we pursued the subject with equal zeal, we could have had the upper hand in space from the start. As it was, we had to embark on what amounted to almost a crash program in space to catch up.

One would think, then, that having had our sensibilities duly jarred once, we would make every effort to see that the Soviets did not get the upper hand in other potentially strategic fields. Yet it is just conceivable that we may end up doing precisely that in parapsychology.

The Russian interest in parapsychology began much as ours did: Russians, too, have these strange experiences that seem not to fit in with known principles and concepts. Just as we in the Western world, they have produced scholars who became fascinated with these experiences and who painstakingly investigated cases. A Commission for the Study of Mental Suggestions, established at the Petersburg Brain Institute by a Russian scholar, V. M. Bekhterev, has probed a number of such reports. Others have been compiled by the late Leonid Leonidovich Vasiliev, a professor of physiology at Leningrad University, who ranks among the Soviet Union's foremost parapsychologists.

Vasiliev himself tells how, at the age of 12, he had a close brush with death which his mother apprehended by what we would now call ESP. Vasiliev recounts how he, his sisters and brothers had been left in the care of several young aunts while their father and ailing mother visited a city some distance away for health reasons. During the course of an imaginative youthful game, Vasiliev recalls, he fell from a willow tree into a river over which the tree extended.

Unable to swim, the hapless Vasiliev began to drown. To parapsychology's everlasting benefit, however, he seized an overhanging branch and pulled himself to the steep bank and safety. "In silent horror," he continues, "my brother and sister were witnessing the event from the tree. We were particularly worried by the inevitability of punishment. We could not conceal from our aunts this adventure: I was completely wet and my brand new high school cap with its white peak—the object of my pride and love—had been carried away by the current."

At home, he says, their young aunts sympathized and agreed not to tell the youngsters' parents.

"They made us promise that we would not repeat it," he recalls. "You can imagine the amazement and confusion—both ours and our aunts'—when, the moment she arrived, our mother described this incident in all details, pointed out the willow, mentioned the cap which had been carried to the dam, etc."

She had dreamed it in detail during her absence and, waking "in tears and disarray," had asked her husband to cable home immediately. "Father admitted that he had not sent a cable but, in order to calm the sick woman, dozed for half an hour in the reception room of the hotel and returned saying he had cabled," Vasiliev relates.

This story and the one that follows appear in Vasiliev's book, *Studies in Mental Telepathy* (Gospolitizidat Publishing House, Moscow, 1962), possibly the best single source of material on early psychical research in Russia.

A teacher, whom Vasiliev identifies as S. Agenosova, tells the following story in a letter dated May 9, 1961, about an apparent telepathic experience between her husband and herself while she was in a Russian city some distance from where her husband was stationed.

"I am able to report to you a seemingly improbable fact. In the autumn of 1942, my husband was attending military school for political personnel in Shadrinsk. I knew that he would be sent to the front after the month of May. I was then employed by the city committee of the association of primary and secondary schools (chairman of the association). Once, in March, I went home quite tired. At that time I lived far from the center and I had to walk the entire way. I remember sitting in a chair in the dining room and falling asleep. Suddenly I saw myself as having received a cable from my husband which read as follows:

"'Leaving Sverdlovsk today, we are going to the front, kisses, Yuriy.' I jumped. There was no cable! However, I had seen it so clearly that I became convinced this cable had to come.

"Forgetting my meal, I immediately went back to the city. I rushed to the city defense section, to its head Leonid Ivanovich Tsypin, and begged him to let me have a pass, telling him that my husband was going to the front and that I had received a cable from him. I took the first train to Sverdlovsk and arrived there at 5 a.m. I had a daughter there whose husband was employed by the NKVD [the Soviet Secret Police]. I went to her home (she is now living with me, Margarita Vasilyevna Skutina), told her of the telegram I had dreamed about. I bought a ticket to Shadrinsk. We had to go to the station and have it punched. We dressed but for about an hour we were unable to leave home. We were sitting in our coats, waiting for something and talking. Suddenly there was a knocking at the door. My daughter opened it. My husband was outside . . . 'Is mother here?' was the first thing he asked. 'She is!' Rita answered. 'I knew it!' It turned out that he had indeed written such a cable but had not sent it, wondering whether I would receive it on time and whether that would only be a useless worry to me. I 'received' this unsent telegram. My husband's train remained in Sverdlovsk two hours and I was able to see him off to the front."

On Dec. 17, 1918, 15-year-old Borya N. Shaber of the city of Vitebsk apparently both saw and heard his young sweetheart at about the time she was dying some distance away. The story was corroborated by no less than six persons to whom the boy had recounted his experience *before he*—or any of the six witnesses—*knew the girl had died*. The witnesses certified their part of the strange case in writing.

But let young Shaber tell his own story:

"On Dec. 17, 1918, at 8:30 a.m. I saw against the wall on which I was resting my legs (I was lying in bed) a round, bright spot which, as I was looking at it, began growing and became the clear outline of a girl. I recognized in her my girl friend Nadezhda Arkad'yevna Nevadovskaya, who was living at that time in Petersburg. Smiling at me, she told me something of which I was able to catch only the last word, 'decay.' After this, the shape of the girl as though walked into the wall and disappeared."

His story, he said, was recorded that same day on paper and signed by six witnesses: S. F. Makunya, I. M. Makunya, A. D. Polesskaya-Shapillo, A. B. Kordukevich, L. Vasil'kovskaya, and A. M. Dombrovskiy.

On Dec. 23, 1918, Shaber said, he received a letter from his girl friend's mother, announcing the girl's death on Dec. 17, 1918. Her last words, according to her mother, were "Borya, there is no dust, there is no decay." Receipt of the letter, plus the nature of its contents, was certified by the above six witnesses.

The sworn statements were placed on file with the Petersburg Brain Institute. The documents were certified by the bookkeeper of the Vitebsk Polytechnical Agricultural Institute, sealed with the seal of the establishment, and signed by two witnesses.

A remarkable story indeed. It appears to be another of those cases in which the receiving person's mind dramatizes, either through a visual or auditory hallucination, a telepathically or clairvoyantly received message.

Collecting stories like this is all very well, of course, but it doesn't constitute "proof" in Russia any more than in the Western world.

At about the time that Dr. Rhine was doing his early work on ESP at Duke University, a leading Soviet physicist, V. F. Mitkevich, was trying telepathy experiments in Leningrad. Instead of using cards like Dr. Rhine, however, Mitkevich used a device similar to a roulette wheel. Each time the whirling wheel came to a stop it left either a black or white screen exposed to the sender's view. The receiver would then try to tell, presumably by telepathy, which color the sender was looking at.

By chance alone, of course, a subject would be right half the time over a long period of trials. Mitkevich did not publish his results; records of the experiment were kept on file with Vasiliev, and, if Vasiliev's description of them is any indication, efforts at statistical analysis were scanty. But in 1934 Vasiliev repeated Mitkevich's work and obtained statistically significant results.

Hypnosis has played a much more prominent role in Russian ESP research than in the United States. If we are to believe the reports, the Russian scientists have had considerable success with the technique. Aided by hypnosis, for instance, Vasiliev actually improved one 24-year-old subject's ESP ability so much that, after 40 tests with that roulette wheel, the subject had achieved 100 per cent accuracy!

Under the conditions of this experiment, Vasiliev attempted to transmit a mental impression of only one color —he chose white—throughout the entire series, to see if the constant repetition would have a conditioning effect on the subject. The subject was in another room throughout the experiment so there was no possibility of an information leak through his normal senses.

The results, as reported by Vasiliev, show a consistency and predictability seldom seen in formal ESP experiments. On the first 10 tests the subject got only four right—less than would be expected by chance. On the second 10 he hit chance squarely on the head—he got exactly five right and five wrong. Then his ESP began to assert itself. On the third test he got eight hits, and on the fourth he achieved a perfect 10 out of 10!

Was he actually learning to recognize the white surface —learning to respond to it over and over by ESP? Or had he, by that mysterious extrasensory process, somehow picked up the fact that the experimenter had decided to concentrate only on white?

In an even more remarkable sequel, Vasiliev switched colors—after the subject had gotten the "white habit," Vasiliev began concentrating only on black. As if carried along by his own momentum, the hypnotized subject continued to name white three more times. Then apparently aware something was amiss, he started naming the black. Seemingly unsure of himself, he switched over to white again. Significantly, Vasiliev says, in the usual experiments where the sender frequently switched colors, neither this subject nor any other named the same color more than three times in a row.

Was this a forerunner of that remarkable series of experiments Milan Ryzl carried out in Czechoslovakia with his librarian subject, Pavel Stepanek? Ryzl, too, you remember (Chapter VI), "trained" his subject with hypnosis; and his subject, too, got consistently high scores identifying colors.

Those pioneering Russian researchers apparently got some wild effects with their hypnosis. Vasiliev tells how he witnessed twin experiments that seemingly fouled each other up—almost as if a pair of telephone wires had gotten crossed. The two experimenters were both trying to transmit telepathic impressions to their own subjects. Each experimenter was in a separate room with his subject. One of the researchers, Dr. N. A. Panov, had his subject deeply hypnotized; the other, Dr. V. I. Rabinovitch, had allowed his to remain awake.

Dr. Panov concentrated on a neutral, apparently innocuous target. His subject received nothing. Meanwhile Dr. Rabinovitch was concocting a wild, emotional fantasy in which he imagined himself watching helplessly while his brother drowned.

Dr. Rabinovitch's subject never got the message. At that precise moment, however, the hypnotized subject of Dr. Panov, to the consternation of the good doctor, began showing visible signs of agitation for no apparent reason. When Dr. Panov asked him what was wrong, the subject described in great detail the dramatic fantasy taking place in Dr. Rabinovitch's head!

In another experiment, Dr. V. N. Finne of Leningrad hypnotized a 29-year-old woman who was suffering from a hysteria-induced paralysis of the left side. Given appropriate suggestions, the patient regained her capacity to move her otherwise useless limbs—not surprising in a case of hysterical paralysis. What *was* surprising was the patient's amazing ability to duplicate whatever motions Dr. Finne executed—without even seeing him. More remarkable still, the patient was apparently able to carry out any designated body movement the experimenter concentrated his *thoughts* on. In a controlled study of this patient's unusual ability, Dr. Finne, and Vasiliev, arranged her room so as to rule out any possible sensory clues from the experimenters. Then one of the experimenters wrote down a brief description of a motion he chose to have communicated to the subject. He handed the written notation to the other experimenter, who tried to "send" it to her telepathically by visualizing the subject carrying out the designated movement. According to Vasiliev, the subject executed each telepathically suggested movement unhesitatingly. Different senders were used, apparently with equal success. Whenever the subject was asked why she had made a given motion, she replied she was "ordered to by . . ."—and she correctly identified the sender, whoever he happened to be.

Reports of the experiment came to the attention of a noted Russian physiologist, Professor A. A. Kulyabko.

Professor Kulyabko's interest was aroused—so much so that he made a special trip to Leningrad for a first-hand demonstration. Here is an English translation of part of the report of that demonstration, taken from a written description in Vasiliev's files and reported in his aforementioned book:

"Suggestion by Professor A. A. Kulyabko, present at the experiment: 'Scratch your left cheek and bridge of the nose'; that was the [mental] suggestion he made sitting on the stool by the head of the bed. In the course of the experiment, the tester repeatedly raised his own right hand and scratched his left cheek. The subject bent her right leg at the knee. She scratched with her right hand her left cheek and lips. She used the same hand to scratch her right cheek."

Although twelve people were present, the subject correctly identified Professor Kulyabko as the originator of the suggestion. When Dr. Finne asked her what Professor Kulyabko had suggested she do, the woman replied: "He terribly irritated the right side of my face."

If this experiment were conducted in a modern parapsychology laboratory, of course, much more stringent precautions would be taken. Films of the subject would be made, and independent judges would be asked to try to match the subjects' photographed movements with the experimenter's previously written descriptions. This would rule out any bias on the part of the observers and the results could be analyzed statistically.

There's a fairly simple experiment, also involving hypnosis, of which the Russians have made great use. In effect, the hypnotist trains his subject to fall asleep whenever the subject is given a telepathic suggestion to do so. The French physiologist Charles Richet reported considerable success with this technique. In Russia, Professor K. I. Platonov, a noted physiologist and hypnotist, evidently made quite a specialty of this. But he had no luck with the method when he simply concentrated on the words "Fall asleep." To be successful, he said, he had visually to imagine his subject sleeping.

It's exceedingly easy to fake this effect, of course. You just hypnotize your subject ahead of time, then tell him that, in the future, he'll fall asleep whenever you give him any of a half-dozen little signals. You can have him fall asleep whenever you raise your little finger, cross your arms, put your hands in your pockets, or lick your lips. It's the oldest kind of stage magic going. But a Soviet psychiatrist, Dr. K. D. Kotkov, refined Platonov's methods and reported success in hypnotizing a subject who was in a separate room and, later, in another part of the city. Precise records were kept of the exact times the subject fell asleep and of the times Dr. Kotkov attempted to induce sleep telepathically.

Vasiliev himself carried out a series of similar experiments. In one, he used two subjects—and found he could selectively induce sleep, telepathically, in either person at will! In another, he carried out some 260 separate trials, telepathically commanding his subjects to go to sleep and, later, to wake up. The statistics are impressive. Only six of the going-to-sleep trials failed! Vasiliev wasn't quite as successful at telepathically waking his subjects up—21 of those attempts failed. But, bear in mind, this was out of a total of 260 trials. You don't need a statistical analysis to see the tremendous significance of these results.

Vasiliev took far greater pains than had his predecessors to rule out any possible normal physical influences. The hypnotist and subject occupied separate metal cabinets, in separate rooms, which screened out electromagnetic effects within the ranges of the medium, short, and ultrashort wavelengths. Long waves were effectively weakened. Yet the cabinets had no influence on the subjects' performance.

These are fairly typical of the kinds of research done up until World War II. Then those horror-torn years of conflict shrouded all pure investigative research in Russia. The immediate postwar years, with Stalin's iron hand held firmly across the face of the nation, were no better. Psychical research during those years was virtually nonexistent.

Then, about 1950, a new openness toward psychical research started to emerge. Abetted by Stalin's death, the newfound tolerance grew. Seemingly, telepathic experiences became fit subjects for discussion.

Parapsychology in the Western world stirred the interest of those Russians who heard about it. What was this strange psychical force that could apparently defy space and time and make mincemeat out of barriers that could stop electromagnetic waves? This kind of thing is profoundly disturbing to Soviet psychology. It is distressingly like something the Russian people used to call Mind or, worse, Spirit, back before the revolution. But the Russians aren't arguing about that today. It's not a case of mind versus matter. If telepathy exists at all, the Russians say, it simply has to have a materialistic explanation. The question is, shall the Russians believe in a materialistic telepathy, or shall they reject telepathy as being dangerously spiritual? Many Russians, scientists and nonscientists alike, give telepathy a resounding "Nyet" because they are seriously afraid it might revive superstition and—horrors—religion. No matter how you look at telepathy, they say, you just don't come up with good old respectable materialism. Too many unseemly pies-in-the-sky creep in—like that opiate "life-after-death" idea. No doubt many American behaviorists would agree.

Vasiliev was at great pains to refute this. He devoted pages of commentary to the degenerate views held by the capitalistic (sic) parapsychologists. No, he asserted, telepathy has nothing to do with any spiritual element in man. It need not threaten the cherished Soviet belief that life ends in oblivion.

"In the capitalistic countries," he writes, "these phenomena are frequently used as a weighty argument supporting the superstitious ideas of the soul, of the 'power of mind over matter.' In our country, the interest displayed by broad population circles in suggestion at a distance is based on phenomena of spontaneous telepathy which take place in daily life. Such phenomena occur in our days among Soviet people who do not share superstitious religious ideas. These phenomena adamantly require a scientific explanation."

But he ran into trouble. Even while he flailed away at religion, he grudgingly admitted that telepathy's failure to behave like electromagnetism poses problems. Exciting problems, yes. It suggests, he said, the existence of a "still unknown factor," which, he dutifully added, must be an "energy factor by the most highly developed matter—the cerebral matter." We get a hint, in his two concluding sentences, that maybe he wasn't that materialistic after all:

"Therefore, something else, something new, should be sought. It has frequently happened in the history of science that the discovery of new facts unexplained on the basis of available knowledge has led to the discovery of unexpected aspects of life."

Interest is growing among the Russian people. Parapsychology is creeping into news reports and magazine articles. Vasiliev's *Mysterious Phenomena of the Human Psyche* has gone through at least three printings. The respected *Literaturnaya Gazeta* featured a discussion between a skeptic, A. Kitaigorodskii, and a believer, a psychiatrist, A. Roshchin. Professor V. P. Tugarinov, head of

Leningrad University's Psychiatry Department, has publicly warned of the dangers of dogmatism in science in this regard. And astronomer F. Zigel has called for a "whole-state approach" to parapsychology.

This has been matched by new interest among some members of the Soviet scientific community. In 1960 Vasiliev established his special laboratory for telepathy research in Leningrad University's Physiology Department. Since his death, the laboratory has continued its work under one of his pupils, Professor P. I. Gulyaev. A second research center has sprung up around a group of Moscow scientists headed by Dr. Alexander B. Kogan, a specialist in cybernetics, within the Popov Scientific and Technical Society for Radiotechnics and Electrocommunication. Formalized into a "bioinformation section," the center has carried out research under its deputy director, E. Naumov, including a series of attempts to send telepathic images more than 2,000 miles. At least one of these efforts, in which the telepathic percipient was identified as one Carl Nikolaiev, was reported as an outstanding success.

What does it all mean for the United States? Could parapsychology, perhaps, provide a common area of research for Soviet and American scientists? Could it perhaps provide grounds for a common view of man's nature—a view which, if ESP and PK are ever understood, both countries could accept?

Or will its potential military applications lead to tight security restrictions? As soon as the Soviet Union gets interested in anything of potential military value—and ESP obviously has such potential—on go the secrecy lids. We can be reasonably certain that, if the Russians think they're onto something that'll give them control over ESP, they won't prance over and announce it to us. Not, at least, until they've actually achieved that control, and then it may be too late.

For years, the notion of space travel was too wild and science fictionish for many Americans to stomach. Then along came Sputnik, which was even tougher to stomach. But we had to do it. Long years of money and effort were needed to land a man on the moon before Russia.

Are we going to go through it all again? American scientists are widely believed to be less receptive to parapsychology than scientists from any other country. The English, the French, the Germans, the Dutch, the Scandinavians, the Japanese—all are more open to it than we are. This is not a good position for a major world power to be in. Except for scattered research at a few institutions—a fair amount of which has been summarized in these pages —American scientists are leaving the field wide open for virtually everyone else.

We have no indication that the Russians are ahead of us in parapsychology. Their politically based philosophy of materialism may yet prove to be as big a block to progress as our academically based materialism. But this is certainly no justification for sitting on our hands. Of such overconfidence was Sputnik born.

* XI *

Other Science Writers View ESP

"I won't have anything to do with those clowns!" one science writer told me.

But another said, "Parapsychology has attracted reputable researchers. It's a legitimate field of inquiry. No science writer should dismiss it."

They were among some 21 science writers I informally questioned on ESP in conversations at a gathering in New Orleans in March, 1969. My aim was to find out how other members of my profession felt about including parapsychology in the science writer's domain, along with physics, chemistry, and biology.

I asked each one whether he thought serious ESP and PK research should be covered by a science writer, by someone else on the paper, or not at all.

I expected—and got—a big variety of answers. Some of those replies hold a promise of better reporting of parapsychology—and hence greater public understanding of it —in the future. Others do not.

The reporters I questioned: Alton Blakeslee, Associated Press; Sandra Blakeslee, *New York Times;* Mark Bloom, *New York Daily News;* Herbert Black, *Boston Globe;*

David Cleary, *Philadelphia Bulletin;* Lewis Cope, *Minneapolis Tribune;* Norbert Dernbach, *Brookhaven Bulletin,* Brookhaven National Laboratory; Donald Drake, *Philadelphia Inquirer;* Bryant Evans, *San Diego Union;* James Hazelwood, *Oakland Tribune;* William Hines, *Chicago Sun-Times;* Miriam Kass, *Houston Post;* Fraser Kent, *Cleveland Plain Dealer;* Jack Martin, *St. Louis Globe-Democrat;* Harry Nelson, *Los Angeles Times;* Judith Randall, *Washington Star;* Delos Smith, United Press International; Arthur Snider, *Chicago Daily News;* Mildred Spencer, *Buffalo Evening News;* Ann Sullivan, *Portland Oregonian;* and Earl Ubell, WCBS-TV, New York.

These people have made careers of reporting news of science and medicine to the public. Almost daily they decide which of the many developments in science are worth reporting and which are not. If they had a chance to choose between covering a meeting on late developments in astronomy, say, or on ESP, which would they choose? Or should science writers be concerned with ESP at all? Their answers, I felt, would tell something about the kind of reporting the public should look for in parapsychology. Their replies would also, perhaps, have a tendency to lend a bit more balance to this book, which, after all, has given the views of only one science writer—myself.

Two science writers out of every three I questioned said parapsychology should be covered by science writers. Commented Harry Nelson of the *Los Angeles Times:*

"I consider it a legitimate field of inquiry. I'm not sold on it by any means, but I think it's something we have to watch closely."

Some were openly enthusiastic. Typical comments:

"It should definitely be part of a science writer's beat. There are just too many unanswered questions—which makes it a subject of legitimate pursuit."

"Yes. I think Dr. Rhine is a serious investigator and this [field] should not be ignored."

"Why sure! A science writer should be a reporter first, and interested in everything."

"Parapsychology is being investigated by some fairly respectable people. I don't think we as reporters are in a position to say it's not science."

"Definitely, yes!"

"I would certainly say it was a legitimate subject of writing and discussion."

Other Science Writers View ESP

"A dean at our local university suggested parapsychology as a subject for our science page. We're going to be doing more with it."

"I feel there's a definite scientific attempt and that people are interested. That makes it a job for a science writer."

"I think there's a definite subject here for a science writer. I wish our paper did more with it. There's a certain amount of resistance, or lack of interest anyway, from higher up."

Some science writers, like the following, expressed confusion:

"I can't dismiss it, yet I've never been sold on it. I don't know of any work that's conclusive. If the city editor asked me to go cover an ESP story, I'd say, 'What are you talking about?' But if scientists held a serious meeting on parapsychology in our city I'd probably cover it the same as everything else."

Caution was voiced by Earl Ubell, former science editor of the *New York Herald Tribune* and subsequently science newscaster for New York's WCBS-TV:

"I think parapsychology should be covered by science writers because we're better qualified to evaluate the material than other reporters. But I'd look at the data very carefully and very skeptically."

Some felt that the science writer's own time should decide the issue. Listen to Art Snider, veteran science writer for the *Chicago Daily News*:

"All of us have limited time. I haven't given much time to ESP. In my very limited examination of the subject I haven't been impressed. I don't want to knock it, you understand. I know there are very reputable people studying it. Yet I don't feel there's enough gold there for me to mine. In cases where Chicago has had meetings and speakers on ESP, we haven't assigned a science writer to cover it—we've assigned other reporters."

Many of the strongest, most emotional statements came from those who were opposed to covering it. "A science writer looks at ESP and forgets it!" quipped one. "I work for a paper that has a lot of crank readers, in a town that's filled with cranks. I don't write about UFO's, I don't write about astrology, I don't write about ESP. If I did, the crank callers would drive me up the walls. We have a limited amount of space, and I don't think we should devote it to something as nebulous as this."

But this reporter added:

"This doesn't mean there's no such thing as ESP. I don't believe in reincarnation or telling the future or God, but none of these precludes ESP. We don't even know how the mind thinks. There's much we still have to learn." (Apparently he didn't know that "telling the future" is part of ESP!)

Said another science writer: "To me, the concept of ESP is unnecessary. We don't need it to help us understand the universe. I rejected God for the same reason."

And another: "I don't think parapsychology should be covered by a science writer, and I don't think psychology or the other social sciences should either. They're not part of science. But that doesn't mean ESP won't be accepted some day. Just because scientists reject an idea now doesn't mean they won't accept it eventually."

Still another commented: "I don't think we should touch it. ESP may happen, but it's important that we retain our credibility."

Dr. Rhine himself is probably better qualified than anyone else to judge the coverage given parapsychology by the press, radio, and TV. He watched it gain the public eye over the years. He, more than anyone else, got the public to think of ESP and PK in terms of scientific research.

"From my experience of more than 40 years," he told me in an interview in March, 1969, "I have the impression that the best coverage was at the time of the first reports of the Duke research on ESP. Science writers and editors under the leadership of Waldemar Kaempffert of *The New York Times* and Dr. E. E. Free of the *Herald Tribune* kept the American public well informed with the progress of research at the laboratory at Duke and at other centers where it was taken up."

Magazines soon picked up the parapsychology story, Dr. Rhine recalled, and radio followed. He described the reporting then as "clear straightforward presentation of findings in laymen's language but no great amount of sensational exaggeration."

Dr. Rhine continued:

"I am satisfied that parapsychology owed much of its support and acceptance to the science writing coverage of the late 1930's and 1940's, but I regret that it was not confined entirely to the professional science writer. Much harm was done, not only by the sensational coverage given by radio programs, but by some of the magazine articles that played up the wrong angles such as the research personalities and areas on the fringe of the research field that were not yet adequately investigated."

The public, he said, became confused. Other scientists became angry, he recalled—especially scientists in related fields such as psychology. They suspected, Dr. Rhine said, that "parapsychologists themselves were responsible for the excessive publicity and more extreme claims."

He went on: "In trying to hold back on releases of findings we were informed by the University Office of Public Information that we were stirring up resentment among the editors and writers and we would get a bad press as a result. The university administration heartily approved of the public interest and we had no choice but to follow the general practice of the campus."

As more popular literature on parapsychology appeared, Dr. Rhine said, American psychologists—whose initial reaction to parapsychology "for the most part was not unfriendly"—became more outspokenly critical. Eventually they stopped commenting on the subject entirely. This "silent treatment" was matched by what Dr. Rhine calls "a similar cultivated indifference on the part of the science writing profession as the new generation came on."

The present generation of science writers, Dr. Rhine believes, tends to judge news according to what our academic institutions say is "good"—at least where science is concerned.

"Is this the way we want it?" he asked. "Is this the sound line to follow for science writing? There are pros and cons, perhaps some even that I do not see or know. As for me, I like to think of the good reporter in science or elsewhere as one who will try to see all he can of what is available and to present as much of it as he can that will interest his readers. The good reporter will not want to be partisan on the new issues that will rise from time to time in a growing field of science. He will want his readers to have the chance to make up their own minds. He will want to be the servant both of the innovator and the interested public. This is his own creative role, the opening up of channels for new ideas to new minds, keeping all within the perspective of good science, which is essential sound judgment."

What, really, is happening? Are parapsychologists getting the press coverage they deserve?

Parapsychology is too often left to reporters who are

not qualified to handle the material. It often gets lumped in with wholly unrelated subjects, such as UFO's and astrology; its most far-out aspects get emphasized. Science writers, judging the subject partly from reports in their own media, tend to rate it below space, biology, and the other sciences.

Parapsychology probably will not get the objective coverage it deserves until science writers give it a higher priority. Yet science writers cannot be expected to do this until more scientists undertake serious research in the field —research that they are willing to talk about publicly. Science writers, like other newsmen, "go where the action is."

* XII *

Summing Up

Until the year 1800, most of the world's respected scientific authorities did not believe in meteorites. Stories of "rocks that fell from the sky" tended to come primarily from the mouths of people who lived in rural areas, where such phenomena were seen most often. They were simple folk, these country-dwellers—not the kind to impress a highly trained scholar. If they said blazing stones sometimes fell from the skies—well, what could you expect from a peasant? They weren't university-educated. They weren't trained thinkers. They didn't realize there were no stones up in the sky.

One of these alleged "rocks from the sky" came to the attention of the renowned Antoine Lavoisier, a founder of modern chemistry. Lavoisier signed a report to the French Academy of Science in 1772 testifying that he had examined this stone. It was, he stated, only an ordinary rock that had been struck by lightning and melted partly into glass.

Yet today we accept the reality of meteorites without question.

Many of the phenomena of parapsychology may be in much the same stage of scientific acceptance—or non-acceptance—as meteorites were when Lavoisier signed that report. In the minds of many educated people, parapsychology is closely linked with superstition and with simple, untrained folk—although this is less true now than it once was. Like meteorites, parapsychological events are hard to catch in the act. It seems to be almost as difficult to get a repeatable experiment in parapsychology as it would be to get a repeatable experiment demonstrating that meteorites do indeed fall from the sky—you have to be there when the action happens.

And yet as new, fresh theories are introduced into psychology, and as more and more daring scientists set up original experiments to explore man's unknown depths, at least some of parapsychology's data may take a place beside the data of biology and physics.

So far, the purely physical sciences have studied only the material components of human life—our chemistry, anatomy, and physiology. And for the most part, the social sciences have studied only gross behavior. Yet we know there is much, much more to a man than this. The human self may be considered a spectrum. On its nearer side, this self is composed of atoms and molecules, of chemical and electrical events, of blood, bone, and muscle. But on its farther side, it is composed of love and hate, of appreciated beauty, of reverence and regret, of values and volition—in short, of the whole complex of states, conscious and unconscious, we call the mind. This farther side of the spectrum has defied most of the instruments of science—hence the tendency of many scientists to dismiss it from consideration. Yet it is precisely this farther side that most concerns us in our dealings with other people. When I tell a joke, for instance, I am not telling it primarily to effect chemical and electrical changes in an organism—indeed, I am not even sure what chemical and electrical changes take place. Nor am I telling it, primarily, to produce a mere "laugh response," as the behaviorists would say—I would soon tire of telling jokes to a computer programmed to laugh at my humor. No, I tell jokes to produce the states of mind which the laughter symbolizes. I tell them to influence the farther side of the spectrum.

This simple example applies to nearly all human relationships. When we deal with people, we are usually trying to do more than influence the material, near side of the spectrum. When we seek the applause of our fellows, we are not striving for the mechanical response of clapping and cheering. Clapping and cheering are only symbols of something deeper—approval and respect, perhaps. We do not write, compose, paint, and make speeches merely to alter the measurable near side of the spectrum; we do these things to affect the states of consciousness that comprise the farther side.

For most of us, this farther side is what ultimately concerns us in other people. It is this, and this alone, that has true intrinsic value. This, and this alone, can evoke genuinely unselfish behavior. This, and this only, can prompt self-sacrifice. It is the farther side of the spectrum that we do things *for*. It is the farther side of the spectrum to which Albert Schweitzer paid homage in revering life. That, and that alone, is important for its own sake.

A materialistic philosophy downgrades the significance of the farther side, even though the materialists may not consciously wish to do so. It takes that which is most important, that which should ultimately concern us in our fellow beings, and makes it a sort of secondary overlay.

I have described the farther side of the human spectrum as "the whole complex of conscious states we call the mind." Yet we do not know how far the spectrum reaches or what it includes. To say that it ends with the portion we directly experience is like saying the universe ends where it ceases to be visible. It is this unknown end of the spectrum that parapsychology, a frontier science of the mind, is probing. It is in these probings that science and the human spirit may find their ultimate meeting place.

Selected Bibliography

BROAD, C. D. *Lectures in Psychical Research.* New York: Humanities Press, 1963.

DUCASSE, C. J. *A Critical Examination of the Belief in a Life After Death.* Springfield, Illinois: Charles C. Thomas, Publishers, 1961.

DUNNE, J. W. *An Experiment with Time.* London: A. C. Black, 1927.

FODOR, NANDOR. *The Haunted Mind.* New York: Helix Press, 1959.

JUNG, CARL G. *Memories, Dreams, Reflections.* New York: Pantheon Books, Inc., 1963.

HANSEL, C.E.M. *ESP—A Scientific Evaluation.* New York: Charles Scribner's Sons, 1966.

PRATT, J. GAITHER. *Parapsychology: An Insider's View of ESP.* New York: E. P. Dutton, 1966.

RHINE, J. B. *The New World of the Mind.* New York: Wm. Sloane Assoc., 1953.

———, and others. *Extrasensory Perception after Sixty Years.* Boston: Bruce Humphries, 1940, 1966.

———, with PRATT, J. G. *Parapsychology: Frontier Science of the Mind.* Springfield, Illinois: Charles C. Thomas, Publishers, 1957, 1962.

———, with BRIER, ROBERT. *Parapsychology Today.* New York: Citadel Press, 1968.

RHINE, LOUISA E. *ESP in Life and Lab: Tracing Hidden Channels.* New York: Macmillan, 1967.

———. *Hidden Channels of the Mind.* New York: Wm. Sloane Assoc., 1961.

SOAL, F. G., and BATEMAN, F. *Modern Experiments in Telepathy.* New Haven: Yale University Press, 1954.

Spraggett, A. *The Unexplained*. New York: New American Library, 1967.

Stevenson, Ian. *Twenty Cases Suggestive of Reincarnation*. New York: American Society for Psychical Research, 1966.

Stearn, Jess. *The Door to the Future*. Garden City, New York: Doubleday, 1963.

Wittkofski, Joseph. *The Pastoral Use of Hypnotic Technique*. New York: Macmillan, 1961.